Seismic Reflections of Rock Properties

Key global geophysical applications, such as prospecting for oil and gas, require interpretation of the seismic response from the subsurface to assess rock properties and conditions behind the observed seismic reflection (amplitude), and to create an accurate geological subsurface model. This book now provides an accessible guide to using the rock-physics-based forward modeling approach, systematically linking the properties and conditions of rock to the seismic amplitude.

Providing a number of practical workflows, the book shows how to methodically vary lithology, porosity, and rock type, as well as pore fluids and reservoir geometry; calculate the corresponding elastic properties; and then generate synthetic seismic traces. These synthetic traces can then be compared to actual seismic traces from the field: the practical implication being that, if the actual seismic response is similar, the rock properties in the subsurface are similar as well. The book catalogues various cases, such as siliclastic and carbonate natural rocks, and time-lapse seismic monitoring, and also discusses the effect of attenuation on seismic reflections. It shows how to build earth models (pseudo-wells) using deterministic as well as statistical approaches, and includes case studies based on real well data.

This is a vital guide for researchers and petroleum geoscientists in industry and academia, providing sample catalogues of synthetic seismic reflections from a variety of realistic reservoir models with direct application to oil and gas exploration and reservoir characterization and monitoring.

Jack Dvorkin is a Senior Research Scientist in the Department of Geophysics at Stanford University. His primary research interests are theoretical rock physics and its practical applications as well as computational rock physics, and he has delivered dozens of industrial rock physics short-courses and lectures worldwide (USA, Canada, Colombia, Brazil, India, China, Japan, Norway, Germany, Italy). Dr. Dvorkin has published around 150 professional papers and has also co-authored two books including *The Rock Physics Handbook* (Cambridge, 2009).

Mario A. Gutierrez is a Principal Geophysicist at Shell International Exploration and Production Inc., working primarily on the application of seismic- and rock physics-based methods for evaluating and risking the presence of reservoir rocks and hydrocarbons to support business decisions and recommendations on oil and gas exploration projects worldwide. Dr. Gutierrez holds a Ph.D. in Geophysics from Stanford University, and has previously held leading applied research and operations positions at Shell, BHP Billiton Petroleum, Ecopetrol, and various seismic contractors, working on rock physics and seismic attributes modeling, reservoir characterization, shallow geo-hazards, and pore pressure prediction.

Dario Grana is an Assistant Professor at the University of Wyoming. He has worked for four years on seismic reservoir characterization at Eni Exploration and Production in Milan, and then moved to Stanford where he received his Ph.D. in geophysics in 2013 – during which time he also published six peer-reviewed journal papers and presented at several international conferences. Dr. Grana's main research interests are rock physics, seismic reservoir characterization, geostatistics, and inverse problems for reservoir modeling.

Seismic Reflections of Rock Properties

Jack Dvorkin

Stanford University, California

Mario A. Gutierrez

Shell International Exploration and Production, Inc. Texas

Dario Grana

University of Wyoming

CAMBRIDGE
UNIVERSITY PRESS

CAMBRIDGE
UNIVERSITY PRESS

University Printing House, Cambridge CB2 8BS, United Kingdom

Cambridge University Press is part of the University of Cambridge.

It furthers the University's mission by disseminating knowledge in the pursuit of education, learning and research at the highest international levels of excellence.

www.cambridge.org
Information on this title: www.cambridge.org/9780521899192

First published 2014

A catalogue record for this publication is available from the British Library

ISBN 978-0-521-89919-2 Hardback

Cambridge University Press has no responsibility for the persistence or accuracy of URLs for external or third-party internet websites referred to in this publication, and does not guarantee that any content on such websites is, or will remain, accurate or appropriate.

To our families

Contents

Color plates appear between pages 126 and 127.

Preface

Rock physics is the part of geophysics concerned with establishing relations between various properties of rocks. Because the elastic radiation is the main agent that allows us to illuminate the subsurface, the primary emphasis of rock physics has been to relate the elastic properties of rock, including the P- and S-wave velocity, P- and S-wave impedance, and Poisson's ratio to porosity, lithology, and pore fluid. A significant number of such rock physics models (transforms) have been developed based on experimental data or physical theories or both. These models enable us to forward model the elastic properties of rock as a function of porosity, rock texture (arrangement of grains or cavities at the pore-scale level), mineralogy, and the compressibility of the pore fluid. Because the seismic reflection depends on the contrast of the elastic properties in the subsurface, rock physics helps generate synthetic seismic reflections at an interface between two different rock types, such as a non-reservoir rock cap and petroleum reservoir, as well as at a fluid contact within a reservoir itself (e.g., gas/oil contact and oil/water contact).

However, in field applications, we face an inverse problem whose solution is not unique: how to interpret a seismic event, which is manifested by a reflection amplitude that stands out of the background, in terms of reservoir properties and conditions. The inherent difficulty of this problem is that the number of variables required to produce the elastic properties of porous rock is larger than the number of seismic observables. Various approaches have been developed to tackle this uncertainty; most of them based on statistical techniques that help assess the probability of the occurrence of a certain object (e.g., high-porosity sand filled with oil) at a given location in a 3D subsurface. Even when using statistical techniques, forward modeling of seismic reflections based on either well log data or theoretical rock physics is a key element of interpreting seismic data.

This book concentrates on such rock-physics-based modeling. The main idea is to create an earth model consistent with the geological understanding at the location, compute the respective elastic properties, and generate synthetic seismic traces. Rock physics enables us to systematically perturb this earth model by varying porosity, mineralogy, and pore fluid, as well as the thickness of the target and the properties of the bounding non-reservoir rock. Systematically conducted perturbational modeling leads to a catalogue of seismic reflections that can serve as a field guide to interpreting a seismic event observed in the field under the assumption that if a synthetic reflection matches the real one, the rock properties and conditions behind the real event may be the same as used to produce the matching synthetic event. Because of the flexibility of existing rock physics models, we can even bracket the interpretation of an event by finding the extreme cases outside which such an event is impossible under site-specific geologic conditions. Here we systematically explore this deterministic as well as statistics-based synthetic forward-modeling approach to generate seismic reflections of rock properties. All examples are based on seismic traces computed for a 1D earth model with the reflecting waves generated by the incidence waves traveling at a varying angle of incidence to an interface between rock types.

This book includes 7 parts, 19 chapters, and an appendix. Part I deals with the basics needed in rock-physics-based forward modeling. In Chapter 1, we introduce the concept of rock-physics-based forward modeling, discuss the non-uniqueness of this approach, revisit the concept of rock physics transforms, and give an example of a synthetic seismic catalogue.

Chapter 2 reviews theoretical and empirical rock physics models, including the velocity–porosity–mineralogy models as well as methods of fluid substitution, that is, computing the elastic properties of a sample filled with a hypothetical fluid if such properties are measured on the same sample filled with a different fluid.

Chapter 3 discusses rock physics diagnostics, a technique that uses well or laboratory data to find an appropriate theoretical model that mimics and explains these data. This process includes two basic steps: bringing all the samples to the common fluid denominator via fluid substitution and then matching the data points with the model curves.

Part II is dedicated to the principles of synthetic seismic modeling. Chapter 4 introduces quick-look rock physics modeling applets, rock-physics-based displays where the user can quickly generate a synthetic gather at an interface depending on the porosity and mineralogy of the reservoir and non-reservoir rock located above it, as well as fluid type and saturation in the reservoir. Such applets can be fairly easily coded in most programming environments.

Chapter 5 discusses the principles of building a pseudo-well relevant to local geology. Special attention is paid to depositionally consistent inputs and the resulting relations between these inputs, such as relations between clay content and porosity and clay content and irreducible water saturation. Compaction trends examined in this chapter also serve as constraints for the porosity range at a given depth.

Chapter 6 discusses the principles of statistics-based perturbations for pseudo-well generation based on existing well data, including interdependence of basic rock properties, such as porosity, clay content, and water saturation. We also describe a procedure to create pseudo-logs of porosity and other rock properties that honor the vertical trends and the vertical continuity present in the input well data. An example is also given of generating these rock properties in 3D, populating a 3D earth model with the elastic properties and generating the normal and offset reflections.

Part III discusses how to use well data and geology to arrive at earth models and respective seismic amplitude. In Chapter 7, we examine two well datasets, both from clastic sedimentary environments, one where the reservoir sand is unconsolidated and soft and the other where the sand is cemented. We systematically conduct the rock physics diagnostics, establish relevant rock physics models, create rock-physics-based modeling applets for both cases, and provide synthetic seismic gathers for various perturbations of the original well data curves.

Chapter 8 discusses vertical log shapes of lithology and porosity associated with various depositional sequences and events in clastic sediments. We construct relevant pseudo-wells and select appropriate rock physics models to translate the sedimentology-driven variables into the elastic properties. Synthetic seismic gathers generated based on such depositional scenarios can serve as a quick-look catalogue for identifying real seismic events.

Chapter 9 discusses rock physics models and laboratory data for carbonate reservoirs. We construct several pseudo-wells to analyze the effects of the pore fluid, porosity, and mineralogy of the seismic gathers as well as generate pseudo-wells and synthetic seismic traces for two concrete depositional scenarios.

In Chapter 10, we address the changes in seismic reflections due to hydrocarbon production (the time-lapse of 4D seismic). The two main factors that affect the elastic properties of the reservoir are the changes in the pore pressure and hydrocarbon saturation. We show how these two factors interact to contribute to discernible temporal amplitude variations or cancel each other and leave the amplitude practically unchanged.

Part IV is dedicated to practical approaches to frontier exploration. Chapter 11 describes a rock-physics-based practical workflow used in hydrocarbon exploration, including selection of a rock physics model, time-to-depth calibration, and depth compaction curves. These results are then used in synthetic seismic gather generation and hydrocarbon- and depth-driven AVO classification for potential hydrocarbon detection.

The following Chapter 12, continues the practical workflow topic and discusses methods of validation of apparent hydrocarbon indicators and assesses the probability of success in actually tapping into a hydrocarbon reservoir after drilling a wildcat well. The direct hydrocarbon indicator check list described in this chapter is placed in the appendix.

Part V deals with advanced rock physics applications. In Chapter 13 we discuss four rock physics diagnostics case studies that reveal the universality of diagenetic trends, meaning that the same theoretical model can be used at different geographical locations and depth intervals; the self-similarity in rock physics, showing that although the porosity and clay content affect the elastic moduli of the rock in separate ways, a certain combination of these two variables can uniquely define that elastic property; a study that reveals that sometimes there is a relation between porosity, stiffness of the rock, and its permeability; and stratigraphy-constrained rock physics modeling based on a dataset from a Tertiary fluvial oil field.

In Chapter 14, we explore the issue of the S-wave velocity and Poisson's ratio prediction using rock physics models. The problem addressed is that many effective-medium-based models predict fairly small Poisson's ratio in gas-saturated sand while the actual well data may sometimes comply with this prediction but also show a much larger Poisson's ratio. We discuss the physical reasons for such disparities and, in the end, show how discrepancies in Poisson's ratio prediction affect synthetic seismic gathers and whether such variations affect the interpretation of a seismic anomaly.

Chapter 15 presents theoretical methods for predicting attenuation from rock properties measured in the well and elaborates on the basics of attenuation theory and experimental results. Example synthetic seismic gathers are generated based on well data with and without including attenuation and are compared to each other.

Chapter 16 is dedicated to the rock physics of gas hydrates and discusses real data collected in gas hydrate wells as well as theoretical models that relate the elastic moduli in sediment with gas hydrates to the hydrate fraction in the pore space. It has been observed that the presence of hydrates often results in significant attenuation of seismic energy traveling through the sediment. This effect is theoretically explained and synthetic seismograms are generated with the attenuation taken into account.

In Part VI we discuss how rock physics operations can be applied directly to seismic amplitude and seismically derived impedance. Chapter 17 presents an example of a rock physics operation, fluid substitution, performed directly on the seismic amplitude. We show, by generating synthetic seismic reflections, that in some cases there are fairly straightforward transforms between the amplitude measured at a wet reservoir and that measured at the same reservoir but filled with gas. Such relations may hold in a range of porosity, mineralogy, and thickness variations. Synthetic modeling is followed by a field example where the original full seismic stack was obtained at a wet reservoir and then transformed into that at a hypothetical gas reservoir. The elements of rock physics analysis behind this transform are discussed in detail.

Chapter 18 shows how simple rock physics analysis can serve to delineate the fluid contact and also map porosity based on a seismic impedance inversion section from a North Sea field.

Part VII is dedicated to an evolving rock physics technique, computational rock physics. It contains one chapter, Chapter 19, where we discuss how computational (or digital) rock physics can serve as a source of controlled experimental data for designing rock physics models and trends in a range of spatial scales of measurement. We discuss the disparity between the scale of various controlled experiments and show that the transforms obtained between two rock properties, such as porosity and velocity, at one scale may be stable in a range of spatial scales.

Acknowledgments

All three authors learned about rock physics and learned rock physics itself at the Stanford Rock Physics Laboratory founded more than three decades ago by Amos Nur. Amos has continuously encouraged advancing rock physics theoretically and experimentally, merging it with geology, and eventually making it a practical instrument in exploration and development. This book would not have happened without Amos. Neither would it have happened without Gary Mavko, one of Nur's first students and a leading rock physics scientist and practitioner, domestically and internationally. His knowledge and advice have contributed to shaping and implementing our ideas. Gary has generously provided the software for generating the synthetic seismic traces used throughout this book. His editorial comments have served to improve the manuscript. We also acknowledge our Stanford colleague Tapan Mukerji for help and support, as well as numerous colleagues in the industry for critically discussing ideas and approaches. The computational rock physics chapter has benefited from help of our colleagues at Ingrain, Inc. We thank Elizabeth Diaz for encouragement and help. We also thank Dawn Burgess for crucial editorial help. The authors thank the Society of Exploration Geophysicists for permission to repurpose material from a number of the authors' papers originally published in *Geophysics* and *The Leading Edge*.

Part I

The basics

1 Forward modeling of seismic reflections for rock characterization

1.1 Introduction

Seismic reflections depend on the contrast of the *P*- and *S*-wave velocity and density between strata in the subsurface, while the velocity and density depend on the lithology, porosity, rock texture, pore fluid, and stress. These two links, one between the rock's structure and its elasticity and the other between the elasticity and signal propagation, form the physical foundation of the seismic-based interpretation of rock properties and conditions. One approach to interpreting seismic data for the physical state of rock is forward modeling. Lithology, porosity, stress, pore pressure, and the fluid in the rock, as well as the reservoir geometry, are varied, the corresponding elastic properties are calculated, and then synthetic seismic traces are generated.

These synthetic traces are compared to real seismic data: full gathers; full stacks; and/or angle stacks. The underlying supposition of such interpretation is that if the seismic response is similar, the properties and conditions in the subsurface that give rise to this response are similar as well. Systematically conducted *perturbational forward modeling* helps create a *catalogue* (field guide) of seismic signatures of lithology, porosity, and fluid away from well control and, by so doing, sets realistic expectations for hydrocarbon detection and monitoring and optimizes the selection of seismic attributes in an anticipated depositional setting.

Key to such perturbational forward modeling are rock-physics-based relations between the lithology, mineralogy, texture, porosity, fluid, and stress in a reservoir and surrounding rock and their elastic-wave velocity and density. To this end, our goal is to elaborate on the details of transforming geologically-plausible rock properties and conditions, as well as reservoir and non-reservoir geometries, into synthetic seismic traces and building catalogues of the synthetic *seismic reflections of rock properties*.

A common result of the remote sensing of the subsurface by elastic waves is the acoustic impedance, the product of the *P*-wave velocity and the bulk density. By itself, it is virtually meaningless to the geologist and engineer. Only after it is interpreted in terms of porosity, lithology, fluid, and stress, can it be used to guide reserve estimates and drilling decisions. A basic problem of such interpretation is that one measured variable (e.g., the impedance) depends on several rock properties and conditions, including the total porosity, clay content, fluid compressibility and density, stress, and rock texture. This means that often it is mathematically impossible to uniquely resolve this problem and predict rock properties from a remote seismic experiment. In other words, in any geophysical interpretation we deal with non-uniqueness, that is, the same seismic anomaly may be produced by more than one combination of underlying rock properties.

A way to mitigate this non-uniqueness is to produce a catalogue of seismic signatures of rock properties constrained by the common geologic sense and site-specific knowledge of the subsurface. The main question addressed in this book is: How to systematically produce such a catalogue within a realistic geology- and physics-guided framework?

1.2 Quantifying elastic properties of earth by forward modeling: a primer

The traditional use of seismic data is to obtain a high-fidelity geometry of geobodies, their boundaries, and accompanying structural heterogeneities, such as faults and folds. This makes the geologic interpretation for prospective hydrocarbon accumulations and their risking elements, including migration, traps, reservoirs, and seals, possible. Seismic impedance inversion techniques (e.g., Russell, 1998; Tarantola, 2005; and Sen and Stoffa, 2013) can be used to look *inside* a geobody, by providing volumes of the elastic properties of its interior. One of the established approaches to impedance inversion is the forward modeling of the seismic signatures of an earth model with an assumed spatial distribution of the velocity and density. This optimization process starts with designing an initial elastic earth model which is gradually perturbed to match synthetic seismograms with real data. Once this match is achieved (within a permissible accuracy tolerance) it is assumed that the underlying elastic earth model reflects the reality. The simplest way of assessing how well the synthetic and real traces match is by visual comparison of the main reflection anomalies at the prospective reservoir and in its vicinity. However, to quantify this process and apply it to a large seismic volume, many different inverse problem solution methods have been proposed and implemented (e.g., Tarantola, 2005; Sen and Stoffa, 2013). Let us concentrate on visual comparison since this is the simplest quick-look method of estimating what rock properties and conditions may be behind a seismic anomaly.

The visual trace comparison methodology is illustrated in Figure 1.1, where a real seismic gather with offset is displayed. The dominant frequency in the seismic data is 30 Hz. To match this gather, a simple 1D elastic earth model is created, where a sand layer with the fixed P- and S-wave velocity (V_p and V_s, respectively) and bulk density (ρ_b) is inserted in shale with fixed elastic properties. A synthetic seismic gather is generated by numerically sending a wavelet of specified shape (a Ricker wavelet with

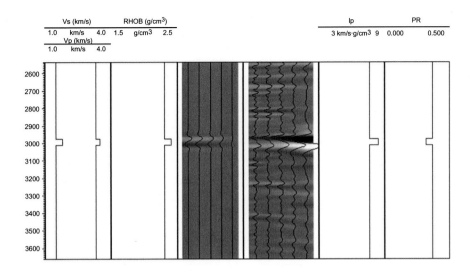

Figure 1.1 Real (fourth track from left) and synthetic (third track from left) seismic gathers. Black is a trough and white is a peak. The elastic earth model used to produce the latter is displayed in the first and second tracks (velocity and density, respectively). The fifth and sixth tracks display the P-wave impedance and Poisson's ratio, respectively. The vertical axis is the true vertical depth (TVD) in meters. The angle of incidence varies from zero to 50 degrees, meaning that the maximum offset is about 3 km. Synthetic gather was generated using a 30 Hz Ricker wavelet. Produced using iMOSS software (Rock Solid Images).

center frequency 30 Hz in this example) through this 1D elastic earth model. Figure 1.1 indicates that the initial guess of the elastic properties of the subsurface did not produce a match between the synthetic and real gather.

Next, we change the elastic properties of the sand layer by reducing its V_p and ρ_b. As a result, the P-wave impedance – $I_p = \rho_b V_p$ – in the sand becomes smaller than that in the background shale. The sand's Poisson's ratio – $v = 0.5(V_p^2 / V_s^2 - 2) / (V_p^2 / V_s^2 - 1)$ – reduces as well. This alteration of the elastic earth model produces a satisfactory match between the synthetic and real seismic gathers (Figure 1.2).

Finally, we vary the elastic properties of *both* the shale and the sand (Figure 1.3) and once again arrive at a satisfactory match between the synthetic and real gathers. This last example highlights the relative nature of the seismic amplitude: the same type of reflection can be produced by more than one set of velocity and density profiles.

Of course, visual comparison of synthetic and real traces is far from being quantitative. Still, it is sufficient for the purpose of this primer. To apply this approach to large seismic volumes, rigorous mathematical methods (e.g., cross-correlation) are employed (e.g., Russell, 1998; Tarantola, 2005; and Sen and Stoffa, 2013).

To further illustrate the relative nature of the seismic amplitude, consider the simplest earth model consisting of two elastic half-spaces. The example in Figure 1.4 shows that the normal reflection is negative as a wave enters the lower half-space where the I_p and v are smaller than those in the upper half-space. The amplitude of

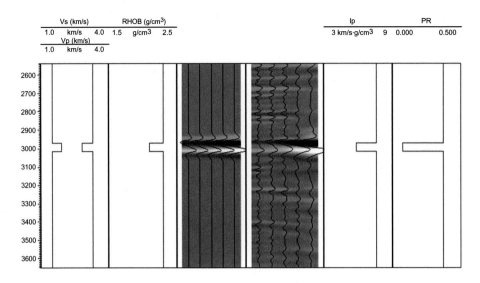

Figure 1.2 Same as Figure 1.1 but with different elastic properties of the sand layer.

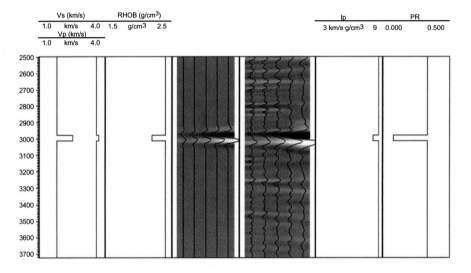

Figure 1.3 Same as Figure 1.2 but with different elastic properties of shale and sand as shown in the first two tracks.

the reflection becomes increasingly negative as the angle of incidence of the wave (or offset) increases.

As we perturb the original earth model by changing the sign of the impedance contrast between the two layers, the synthetic reflections change, both qualitatively and quantitatively (Figure 1.5). As we continue to perturb the elastic properties, we arrive at reflections very similar to those displayed in Figure 1.4 but with a different elastic input (Figure 1.6).

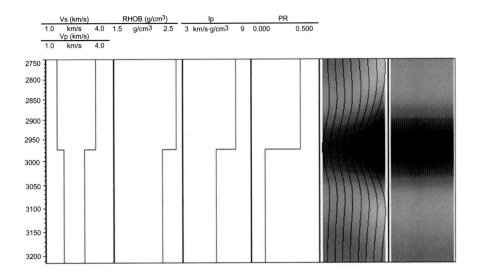

Figure 1.4 Synthetic seismic gather (fifth track) and full-offset stack (sixth track) versus depth (m). The first four tracks show the input elastic properties in this earth model. The angle of incidence varies from zero to 50 degrees. Generated by a 30 Hz Ricker wavelet. Produced using iMOSS software (Rock Solid Images).

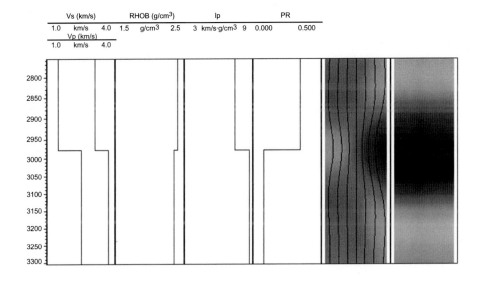

Figure 1.5 Same as Figure 1.4 but with different elastic properties of the layers (as displayed in the left-hand tracks).

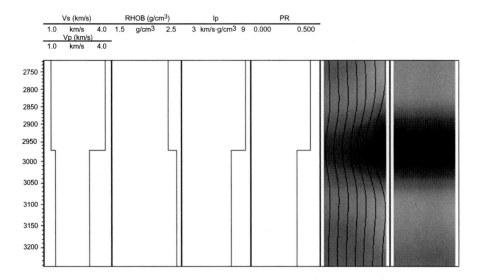

Figure 1.6 Same as Figure 1.4 but with different elastic properties of the layers (as displayed in the left-hand tracks).

This example illustrates the dichotomy in geophysical remote sensing, which is both relative and absolute: while the seismic reflection relates to the impedance contrast, the reservoir properties, such as porosity, relate to the absolute value of the impedance. One way of interpreting the relative in terms of the absolute is to perturb the absolute and calculate the corresponding relative.

Clearly, such interpretation is not unique. Different earth models can produce the same response. In traditional impedance inversion, this non-uniqueness is mitigated by anchoring the elastic properties to a nearby well. Once an absolute impedance volume is available, impedance–porosity, impedance–lithology, and impedance–fluid transforms can be applied to it to map these reservoir properties.

Still, even if a perfect impedance volume of the subsurface is available and appropriate transforms have been established, their application to seismic impedance may not be straightforward because usually such transforms are obtained at the laboratory or well log scale (inch or foot) while seismic impedance maps have the seismic scale which is much larger (hundreds of feet). This means that seismic interpretation for rock properties is never unique. This non-uniqueness comes from at least two sources: (a) the scale disparity between the traditional experiment-based (laboratory and/or well data) rock physics and seismic scales; and (b) the relative versus absolute disparity between the seismic reflection and the actual physical impedance. Yet another source stems from the possibility that the same elastic properties can, in principle, result from various combinations of mineralogy, porosity, and pore fluid (see Chapter 2).

The non-uniqueness can be reduced if geological reasoning is used in reducing the number of variants of an elastic earth model. This can be done by perturbing the fundamental rock properties, such as porosity and mineralogy, calculating the resulting elastic properties, and, finally, using these elastic properties in synthetic seismic generation. Such an approach helps constrain the range of the earth models by selecting porosity and mineralogy in a relatively narrow domain relevant to local geology.

Moreover, such an approach helps construct *what if* scenarios by moving rock in geologic space and time according to the laws of geology. This is why in the next example we perturb the bulk properties and conditions and arrive at a synthetic-to-real seismogram match.

1.3   Quantifying rock properties by forward modeling: a primer

In Figure 1.7 we first produce a synthetic seismogram by making an assumption about the porosity and mineralogy of shale and sand and also assuming that the sand is fully water-saturated. This first attempt at matching the real gather fails (Figure 1.7, top).

Next, we keep the porosity and mineralogy the same but partly replace water with large amounts of gas. As a result, we arrive at a reasonable match between the synthetic and real gather. Finally, we increase the water saturation and reduce the gas saturation accordingly. The resulting synthetic gather still matches the real one.

One conclusion of this exercise is that there are definitely hydrocarbons in the reservoir but their amount cannot be predicted from seismic data. That is, seismic reflections are weakly sensitive to gas saturation and, hence, apparently, in this case do not help discriminate commercial gas volumes from residual gas.

At the heart of this forward-modeling approach is a rock physics transform from porosity, mineralogy, rock texture, and fluid to the elastic properties of rock.

1.4   Rock physics transforms: a primer

One way of obtaining a relevant rock physics transform is by examining data which include the basic rock properties (e.g., porosity and mineralogy) *and* elastic properties measured on the *same samples*. If these data are matched by an existing rock physics model, then this model *is* the transform to be used in synthetic seismic modeling. These data may come from the laboratory or well logs.

An example is shown in Figure 1.8 where laboratory data from a large number of sandstone samples spanning ranges of porosity and clay content (Han, 1986) are matched by the Gal *et al.* (1998) velocity–porosity–mineralogy model.

The low-clay-content outlier at high porosity is unconsolidated Ottawa sand whose texture is different from that of the other competent-sandstone samples. The transform used works for the latter. Clearly, a different transform has to be found for unconsolidated sand.

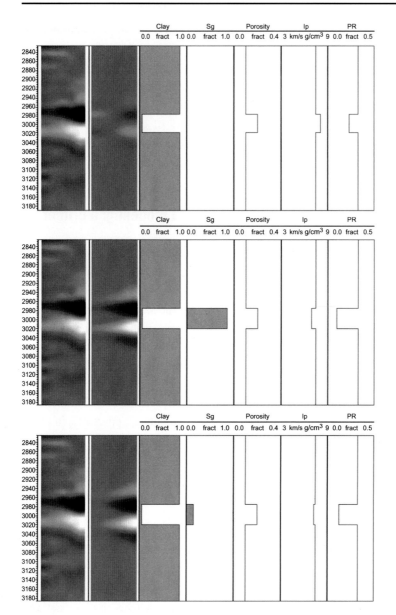

Figure 1.7 From left to right: real seismic gather, the same as displayed in Figure 1.1; synthetic seismic gather; clay content; gas saturation (one minus water saturation); total porosity; and the resulting *P*-wave impedance and Poisson's ratio. Top: Full water saturation. Middle: low water saturation. Bottom: High (but not 100%) water saturation. The angle of incidence varies from zero to 50 degrees. Frequency is 30 Hz. Produced using iMOSS software (Rock Solid Images).

1.5   Synthetic seismic catalogues

An example of a rock-physics-based synthetic seismic catalogue is displayed in Figure 1.9, where we vary the porosity of shale as well as the hydrocarbon saturation

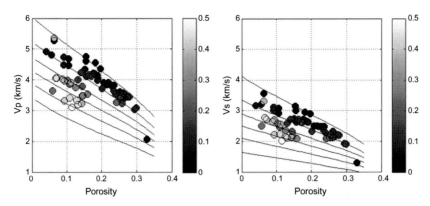

Figure 1.8 Left: The *P*-wave velocity versus porosity. Right: the *S*-wave velocity versus porosity. Symbols: Han's (1986) measurements on room-dry clean and shaly sandstone samples at 40 MPa hydrostatic effective pressure, color-coded by the clay content. Curves: the stiff-sand model (Gal *et al.*, 1998; and Mavko *et al.*, 2009). The upper curve is for zero clay content; the lower curve is for 100% clay content; the curves in the middle are for varying clay content with a 20% increment (from top to bottom).

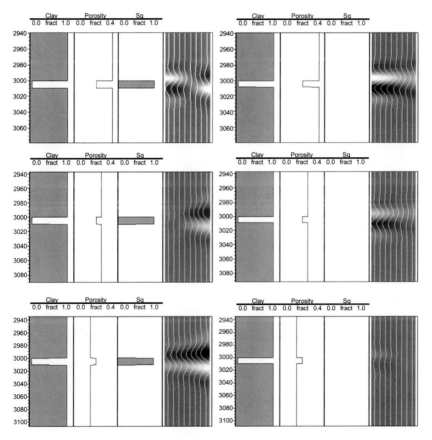

Figure 1.9 Top to bottom: synthetic seismic gathers corresponding to three variants of porosity in shale and sand. Left: Low water saturation. Right: Full water saturation. The first track is depth (m). Following (from left to right) are clay content, porosity, gas saturation, and synthetic seismic gather. The angle of incidence varies from zero to 50 degrees. Frequency is 30 Hz. Produced using iMOSS software (Rock Solid Images).

in the sand to arrive at synthetic seismic gathers. The rock physics transform used here (in both shale and sand) is the soft-sand model appropriate for unconsolidated sediment (Dvorkin and Nur, 1996; Chapter 2).

In the following chapters we will elaborate on rock physics transforms as well as build catalogues based on various geologic scenarios.

2 Rock physics models and transforms

2.1 Rock physics transforms

A meaningful way of designing a transform from one rock attribute (e.g., the elastic wave velocity) to another (e.g., porosity) is to examine data where both attributes are measured on the same sample. Then either an empirical or an analytical model can be constructed to honor and generalize these data. Rock physics transforms are a valuable complement to data because once a relation between various attributes is understood and supported by basic theory, it can be used to extend the range of their applicability beyond the original dataset.

The existing transforms are exhaustively discussed in Avseth *et al.* (2005) and Mavko *et al.* (2009). Here we revisit only the transforms and models directly used in the following discussion.

2.2 Elastic constants

If rock is assumed to be an isotropic linear elastic body, its deformation under stress is governed by two independent elastic moduli, such as the bulk modulus, K, and shear modulus, G. The former is defined as the ratio of hydrostatic stress to volumetric strain, while the latter is the ratio of a shear stress to the resulting shear strain.

The elastic-wave velocities are related to these elastic moduli and the density of the elastic body (ρ). There are two types of waves that can propagate in an isotropic elastic body: the compressional wave whose speed is V_p and the shear wave whose speed is V_s:

$$V_p = \sqrt{\frac{M}{\rho}}, \quad V_s = \sqrt{\frac{G}{\rho}}, \quad M = K + \frac{4}{3}G, \tag{2.1}$$

where M is called the compressional modulus.

All other elastic constants, such as Young's modulus, E, and Poisson's ratio, v, can be derived from K and G:

$$E = \frac{9KG}{3K+G}, \quad v = \frac{3K-2G}{2(3K+G)} = \frac{1}{2}\frac{V_p^2/V_s^2 - 2}{V_p^2/V_s^2 - 1}. \tag{2.2}$$

Two other elastic constants used in geophysics are Lamé's constants, λ and μ:

$$\lambda = K - \frac{2}{3}G, \quad \mu = G. \tag{2.3}$$

Finally, the P-wave (I_p) and S-wave (I_s) impedances are defined as

$$I_p = \rho V_p, \quad I_s = \rho V_s. \tag{2.4}$$

2.3   Solid phase

Essentially all rock physics models dealing with the elastic properties of rock treat it as a composite made of two basic elements: the mineral frame and the pore fluid. The mineral frame often includes more than one mineral. A traditional treatment of this situation is to analytically create a single, or "effective," mineral whose elastic properties depend on those of the mineral constituents, which can be used to calculate the elastic properties of the dry mineral frame.

To accomplish this task, we examine this hypothetical effective mineral (effective solid phase) that is a composite itself, made of several pure-mineral elastic components with known volumetric fractions. There are the upper and lower bounds for the bulk and shear moduli of this elastic composite, the Voigt (K_V and G_V) and Reuss (K_R and G_R) bounds, respectively (e.g., Mavko *et al.*, 2009):

$$K_V = \sum_{i=1}^{N} f_i K_i, \quad G_V = \sum_{i=1}^{N} f_i G_i,$$
$$K_R^{-1} = \sum_{i=1}^{N} f_i K_i^{-1}, \quad G_R^{-1} = \sum_{i=1}^{N} f_i G_i^{-1}, \tag{2.5}$$

where N is the number of mineral components; f_i is the volume fraction of the ith component in the solid phase of the rock (all f_i sum up to 1); and K_i and G_i are the bulk and shear moduli of the ith component. The effective bulk and shear moduli of the effective solid are often computed as the Hill's average of these bounds (e.g., Mavko *et al.*, 2009):

Table 2.1 *Elastic moduli and densities of minerals used in this book.*

Mineral	Bulk Modulus (GPa)	Shear Modulus (GPa)	Density (g/cm^3)
Quartz	36.6	45.0	2.65
Clay	21.0	7.0	2.58
Feldspar	75.6	25.6	2.63
Calcite	76.8	32.0	2.71
Dolomite	94.9	45.0	2.87

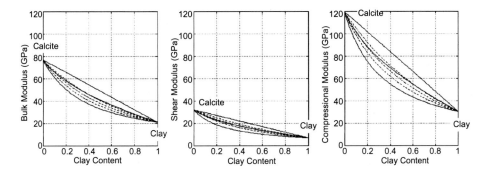

Figure 2.1 Calculated elastic moduli of a clay/calcite mixture plotted versus clay content. From left to right: bulk, shear, and compressional moduli. The upper solid curve is the Voigt (upper) bound. The lower solid curve is the Reuss (lower) bound. The middle solid curve is the Hill average (the arithmetic average of these upper and lower bounds). The upper dashed curve is the upper Hashin–Shtrikman bound. The lower dashed curve is the lower Hashin–Shtrikman bound. The middle dashed curve is the arithmetic average between the Hashin–Shtrikman bounds.

$$K_H = \frac{K_V + K_R}{2}, \quad G_H = \frac{G_V + G_R}{2}. \tag{2.6}$$

The elastic moduli of individual minerals may each span a range of values. We recommend selecting a single set of values for each mineral and then consistently using them in modeling. Mavko *et al.* (2009) provide a detailed table of these properties based on previously published measurements. A subset of these values is listed in Table 2.1 and will be used throughout this book unless stated otherwise.

As an example of this approach, the Reuss and Voigt bounds, as well as Hill's average, are computed for a mixture of two elastically contrasting minerals, soft clay and stiff calcite (Figure 2.1).

A very similar approach can be used with more rigorous elastic bounds, the Hashin–Shtrikman (1963) upper and lower bounds:

$$K_{HSUP} = \left[\sum_{i=1}^{N} \frac{f_i}{K_i + (4/3)G_{max}} \right]^{-1} - \frac{4}{3}G_{max},$$

$$K_{HSLO} = \left[\sum_{i=1}^{N} \frac{f_i}{K_i + (4/3)G_{min}} \right]^{-1} - \frac{4}{3}G_{min},$$

$$G_{HSUP} = \left[\sum_{i=1}^{N} \frac{f_i}{K_i + Z_{max}} \right]^{-1} - Z_{max}, \quad G_{HSLO} = \left[\sum_{i=1}^{N} \frac{f_i}{K_i + Z_{min}} \right]^{-1} - Z_{min}, \tag{2.7}$$

$$Z_{max} = \frac{G_{max}}{6} \frac{9K_{max} + 8G_{max}}{K_{max} + 2G_{max}}, \quad Z_{min} = \frac{G_{min}}{6} \frac{9K_{min} + 8G_{min}}{K_{min} + 2G_{min}},$$

where the subscripts $HSUP$ and $HSLO$ refer to the upper and lower bound, respectively, and max and min refer to the maximum and minimum elastic moduli, respectively, among the mineral constituents. Similar to Hill's average, we can assume that the effective elastic moduli of the composite solid phase (K_{HS} and G_{HS}) are

$$K_{HS} = \frac{K_{HSUP} + K_{HSLO}}{2}, \quad G_{HS} = \frac{G_{HSUP} + G_{HSLO}}{2}. \tag{2.8}$$

These bounds and the corresponding effective elastic moduli for a mixture of clay and calcite are also displayed in Figure 2.1. The values given by Eq. (2.6) are fairly close to those given by Eq. (2.8). In this book we will consistently use Hill's average, as given by Eq. (2.6). The bulk and shear moduli of the effective solid phase will be called K_s and G_s, respectively, while the corresponding compressional modulus and Poisson's ratio will be called M_s and v_s, respectively.

The density of a multi-mineral solid phase (ρ_s) is a volume-weighted arithmetic average of the components (ρ_i):

$$\rho_s = \sum_{i=1}^{N} f_i \rho_i. \tag{2.9}$$

The corresponding P- and S-wave velocities (V_{ps} and V_{ss}, respectively) are

$$V_{ps} = \sqrt{M_s / \rho_s}, \quad V_{ss} = \sqrt{G_s / \rho_s}. \tag{2.10}$$

These velocities for a clay/calcite mixture are displayed in Figure 2.2. The curves based on Hill's average are very close to those based on the Hashin–Shtrikman average.

2.4   Fluid phase

A similar approach is used to produce the bulk modulus of the "effective" fluid phase, the components of which may include water, oil, and gas. If all individual fluid phases

Table 2.2 *Elastic moduli and densities of water, oil, and gas for salinity 40,000 ppm; oil API gravity 30 and GOR 300; gas gravity 0.7; and 20 MPa (2,900 psi) pressure and 60° C (140° F) temperature.*

Fluid	Bulk Modulus (GPa)	Density (g/cm³)
Water	2.6819	1.0194
Oil	0.3922	0.6359
Gas	0.0435	0.1770

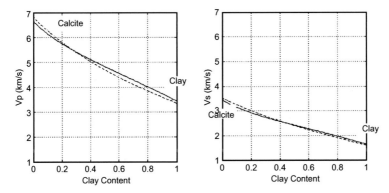

Figure 2.2 Same as Figure 2.1 but for the *P*- (left) and *S*- (right) wave velocity. Only the Hill average (solid) and the arithmetic average for the Hashin–Shtrikman bounds (dashed) are shown.

remain in perfect hydraulic communication, that is, the pressure in the gas is the same as in the oil and the same as in the water, the effective bulk modulus of such an immiscible system (K_f) is

$$\frac{1}{K_f} = \frac{f_w}{K_w} + \frac{f_o}{K_o} + \frac{f_g}{K_g},$$ (2.11)

where *f* is the volume fraction of the phase ($f_w + f_o + f_g = 1$); *K* is the respective bulk modulus; and the subscripts *w*, *o*, and *g* refer to water, oil, and gas, respectively. The effective density is calculated as the weighted arithmetic average, the same as in Eq. (2.9). Since most fluids do not resist shear deformation, the shear modulus is zero.

The properties of pore fluids strongly depend on water salinity, oil gravity and gas content in oil (GOR), gas gravity, and pressure and temperature. These properties can be calculated from the Batzle–Wang (1992) equations (see Section 2.13 below). An example is given in Table 2.2.

The bulk moduli of water/oil and water/gas immiscible systems are plotted versus water saturation S_w (same as f_w) in Figure 2.3. Because the bulk modulus of the system is the harmonic average of those of the components, the softer component dominates

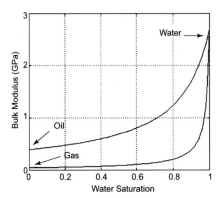

Figure 2.3 Bulk modulus of immiscible water/oil and water/gas systems versus water saturation, according to Eq. (2.11).

and the effective bulk modulus remains close to that of the hydrocarbon until water saturation becomes fairly high.

Brie *et al.* (1995) suggested a different mixing law for fluid phases (see discussion in Mavko *et al.*, 2009) that produces a stiffer effective fluid at the same saturation. This mixing law does not have intuitive physical meaning but the fluid substitution results that use it match some of the well data where patchy saturation likely occurs (see physically consistent theory of patchy saturation in the next section).

2.5   Fluid substitution

The elastic moduli of rock with fluid differ from those of the dry frame. Figure 2.4 illustrates this observation by displaying laboratory high-frequency measurements on a high-porosity unconsolidated sand sample. The difference between the dry-rock and wet-rock bulk modulus is dramatic.

Gassmann (1951) introduced a theoretical fluid-substitution equation that provides the bulk modulus in fluid-saturated rock (K_{Sat}) as a function of the dry-rock bulk modulus (K_{Dry}), the bulk modulus of the solid phase (K_s), that of the pore fluid (K_f), and total porosity (ϕ). Also, the wet-rock shear modulus is the same as the dry-rock one:

$$K_{Sat} = K_s \frac{\phi K_{Dry} - (1+\phi)K_f K_{Dry} / K_s + K_f}{(1-\phi)K_f + \phi K_s - K_f K_{Dry} / K_s}, \quad G_{Sat} = G_{Dry}. \tag{2.12}$$

The assumptions underlying this equation are: (a) the rock is elastically isotropic; (b) the mineral phase of the rock can be characterized by a single bulk modulus and density; (c) the fluid phase in the pore space can also be characterized by a single bulk modulus and density; and (d) fluid in the pores of the rock volume under examination

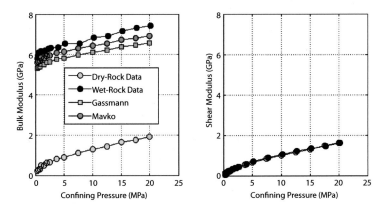

Figure 2.4 Bulk (left) and shear (right) moduli of high-porosity unconsolidated sand versus hydrostatic confining pressure. The pore pressure is constant 0.1 MPa. Filled symbols are data obtained in ultrasonic pulse transmission experiments on the water-saturated sample. Open symbols are for the room-dry sample. Gray squares in the left-hand frame are Gassmann's fluid substitution from the dry-rock data to wet rock. Gray circles are the same fluid substitution but using the Mavko (*P*-wave only) method. Data are from Zimmer (2003).

is in perfect hydraulic communication, that is, any pore pressure disturbance is immediately transferred throughout all pores. To address the first two assumptions in the case of a multi-mineral rock and multiphase pore fluid, we use Eqs (2.8) and (2.11), respectively.

The third assumption implies that Eq. (2.12) is only applicable for low-frequency excitation of rock, such as during seismic wave propagation and, arguably, well sonic and dipole measurements. This is the reason why the wet-rock bulk modulus according to Eq. (2.12) and displayed in Figure 2.4 is smaller than that measured in the laboratory at very high (about 1 MHz) frequency.

A reverse form of Eq. (2.12) offers the dry-rock bulk modulus as a function of measured (low-frequency) wet-rock modulus:

$$K_{Dry} = K_s \frac{1 - (1-\phi)K_{Sat}/K_s - \phi K_{Sat}/K_f}{1 + \phi - \phi K_s/K_f - K_{Sat}/K_s}, \quad G_{Dry} = G_{Sat}. \qquad (2.13)$$

Equations (2.12) and (2.13) comprise a method for computing the bulk modulus of rock saturated with effective fluid, B, that has bulk modulus K_{fB} if the bulk modulus of the same rock but saturated with fluid A with bulk modulus K_{fA} is known to be K_{SatA}:

First we have to compute the dry-rock bulk modulus using Eq. (2.13) as

$$K_{Dry} = K_s \frac{1 - (1-\phi)K_{SatA}/K_s - \phi K_{SatA}/K_{fA}}{1 + \phi - \phi K_s/K_{fA} - K_{SatA}/K_s}. \qquad (2.14)$$

Then we use Eq. (2.12) to compute the bulk modulus of the rock saturated with fluid B as

$$K_{SatB} = K_s \frac{\phi K_{Dry} - (1 + \phi)K_{fB}K_{Dry} / K_s + K_{fB}}{(1 - \phi)K_{fB} + \phi K_s - K_{fB}K_{Dry} / K_s}. \tag{2.15}$$

The shear modulus remains the same:

$$G_{SatB} = G_{SatA} = G_{Dry}. \tag{2.16}$$

The bulk density of the rock, ρ_b, also changes as the pore fluid changes:

$$\rho_{bB} = \rho_{bA} - \phi\rho_{fA} + \phi\rho_{fB}, \tag{2.17}$$

where ρ_{bA} and ρ_{bB} are the bulk densities of the rock with fluid A and B, respectively, and ρ_{fA} and ρ_{fB} are the densities of fluids A and B, respectively.

Finally, the elastic-wave velocity for the rock saturated with fluid B is computed using Eqs (2.1). Note that the change will affect not only V_p but also V_s, the latter due to the change in the bulk density of the rock.

The Gassmann fluid substitution implies that the bulk modulus is known, that is, both V_p and V_s are available:

$$K = \rho V_p^2 - \frac{4}{3}\rho V_s^2. \tag{2.18}$$

A problem with this fluid substitution method arises where the shear-wave data are not available or not reliable. Mavko *et al.* (1995) offer V_p-only fluid substitution that operates on the compressional modulus instead of the bulk modulus. The only difference between this and Gassmann's fluid substitution is that the compressional moduli, M_{Sat}, M_{Dry}, and M_s, are used instead of the corresponding bulk moduli:

$$M_{Sat} \approx M_s \frac{\phi M_{Dry} - (1 + \phi)K_f M_{Dry} / M_s + K_f}{(1 - \phi)K_f + \phi M_s - K_f M_{Dry} / M_s},$$

$$M_{Dry} \approx M_s \frac{1 - (1 - \phi)M_{Sat} / M_s - \phi M_{Sat} / K_f}{1 + \phi - \phi M_s / K_f - M_{Sat} / M_s}. \tag{2.19}$$

The compressional modulus, either for saturated or dry rock, is the product of V_p squared and the bulk density. The latter, of course, has to be that of the saturated or dry rock, respectively.

The "effective-pore-fluid" rule as expressed by Eq. (2.11) implies that the phases of the pore fluid are in perfect hydraulic communication, that is, pressure fluctuations experienced by, for example, water are immediately translated into the other phase, for

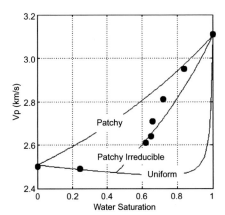

Figure 2.5 *P*-wave velocity versus water saturation. The lower curve is for uniform saturation. The upper curve is for patchy saturation with zero irreducible water saturation and zero residual gas saturation. The intermediate curve is for patchy saturation with irreducible water saturation 0.45 and zero residual gas saturation. The symbols are data adopted from Lebedev *et al.* (2009). The three theoretical curves are computed from the dry-rock data using the Mavko *et al.* (1995) method.

example, gas. Because the softer, more compressible, phase dominates this effective bulk modulus, the latter remains very close to that of the softer phase essentially in the entire range of water saturation (except, of course, at full water-saturation). This situation is called *uniform* saturation. The resulting bulk and compressional moduli will remain close to those of the gas-saturated rock except for a narrow water-saturation range near $S_w = 1$. The *P*-wave velocity outside this narrow range will be even smaller than that of the gas-saturated rock due to the increase in the bulk density as water is added to the pores (Figure 2.5).

However, well and laboratory (Figure 2.5) data sometimes show departure from a theoretical uniform-saturation curve. A common assumption used to explain this observation is that each of the pore-fluid phases is grouped in relatively large patches and the hydraulic communication between these patches is restricted, because of their size, even at relatively low frequencies.

To quantify the effect of the fluid on the elastic properties of rock in this case, assume that the irreducible water saturation is S_{wi} while the residual hydrocarbon saturation is S_{hr}. The bulk moduli of water and hydrocarbon are K_w and K_h, respectively. Also assume that for $S_w \leq S_{wi}$ and $S_w \geq 1 - S_{hr}$ saturation is uniform. Then for $S_{wi} < S_w < 1 - S_{hr}$ the compressional (M_{Patchy}) and bulk (K_{Patchy}) moduli are (Mavko *et al.*, 2009)

$$\frac{1}{M_{Patchy}} = \frac{x}{M_2} + \frac{1-x}{M_1}, \quad K_{Patchy} = M_{Patchy} - \frac{4}{3}G_{Dry}, \tag{2.20}$$

where

$$x = \frac{1 - S_w - S_{hr}}{1 - S_{wi} - S_{hr}};$$

$$M_1 = \frac{4}{3}G_{Dry} + K_s \frac{\phi K_{Dry} - (1 + \phi)K_{f1}K_{Dry} / K_s + K_{f1}}{(1 - \phi)K_{f1} + \phi K_s - K_{f1}K_{Dry} / K_s}, \quad \frac{1}{K_{f1}} = \frac{S_{hr}}{K_h} + \frac{1 - S_{hr}}{K_w}; \tag{2.21}$$

$$M_2 = \frac{4}{3}G_{Dry} + K_s \frac{\phi K_{Dry} - (1 + \phi)K_{f2}K_{Dry} / K_s + K_{f2}}{(1 - \phi)K_{f2} + \phi K_s - K_{f2}K_{Dry} / K_s}, \quad \frac{1}{K_{f2}} = \frac{1 - S_{wi}}{K_h} + \frac{S_{wi}}{K_w}.$$

For the V_p-only fluid substitution method, Eq. (2.19) can be used but with the bulk modulus replaced by the compressional modulus:

$$M_1 = M_s \frac{\phi M_{Dry} - (1 + \phi)K_{f1}M_{Dry} / M_s + K_{f1}}{(1 - \phi)K_{f1} + \phi M_s - K_{f1}M_{Dry} / M_s},$$

$$M_2 = M_s \frac{\phi M_{Dry} - (1 + \phi)K_{f2}M_{Dry} / M_s + K_{f2}}{(1 - \phi)K_{f2} + \phi M_s - K_{f2}M_{Dry} / M_s}. \tag{2.22}$$

The two patchy-saturation V_p curves according to Eq. (2.21) and for $S_{hr} = 0$ and S_{wi} zero and 0.45 are displayed in Figure 2.5. The latter curve captures the experimental results.

Perhaps the strongest assumption in the patchy saturation theory is that a patch is hydraulically disconnected from its neighbors, meaning that the wave-induced pore pressure perturbations do not equilibrate between the patches over the wave period. This means that the physical size of a patch (e.g., its diameter if it is approximately spherical) is frequency dependent and should not be smaller than the diffusion length L, where $L = 2\sqrt{kK_f / (f\eta\phi)}$ and L is in meters, f is the frequency in 1/s, k is the permeability in m^2, K_f is the fluid bulk modulus in Pa, η is the fluid dynamic viscosity in Pa s (1 cPs = 10^{-3} Pa s), and ϕ is the porosity.

At the same time, we assume that these pore pressure perturbations *do* equilibrate within each patch. This means that the patch size should not be much larger than the diffusion length. In practical applications of this theory we select the subsample size equal to the average diffusion length for all the subsamples. See more discussion in Chapter 10.

2.6   The Raymer–Hunt–Gardner transform

The classical Raymer–Hunt–Gardner (1980) functional form for estimating V_p in porous rock with fluid from porosity, ϕ, and the P-wave velocity in the mineral and fluid phases, V_{ps} and V_{pf}, respectively, is

$$V_p = (1 - \phi)^2 V_{ps} + \phi V_{pf}, \tag{2.23}$$

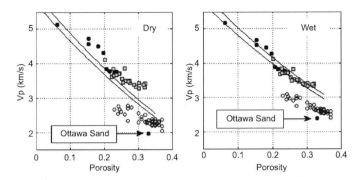

Figure 2.6 *P*-wave velocity versus porosity. Left: room-dry data measured at 30 MPa confining pressure. Right: Wet-rock data obtained from the room-dry data using Gassmann's fluid substitution from air to water with bulk modulus 2.25 GPa and density 1 g/cm³. Black symbols are from Han's (1986) dataset for clay content below 5%; gray squares are from the Strandenes (1991) dataset; and gray circles are from Blangy's (1992) dataset. The arrows point to the Ottawa datapoint included in Han's (1986) dataset. The upper curve is from RHG for 100% quartz content while the lower curve is for 95% quartz and 5% clay content.

where V_{ps} and V_{pf} can be obtained from Eqs (2.8) and (2.11), respectively, and the densities of these phases. An abbreviation for this model used throughout this book is RHG.

In Figure 2.6 we compare the RHG predictions with laboratory data obtained on three different sets of relatively clean (clay-free) sandstones. The measurements used here were obtained on room-dry samples at 30 MPa confining pressure. We bring these data to the wet-rock conditions by applying Gassmann's fluid substitution to these dry-rock data. This comparison shows that (a) RHG accurately reproduces the wet-rock data but somewhat underestimates the dry-rock data and (b) RHG only works for "fast" (or competent) sediment and strongly overestimates the velocity in unconsolidated friable sand. Both these conclusions are expected, since the RHG model is an empirical one and was established based on water-saturated competent-rock data which did not include friable sands.

Specifically, the wet-rock RHG curves reproduce the trend formed by clean samples from two datasets, one by Han (1986) and the other by Strandenes (1991). The former includes consolidated low to medium-porosity samples while the latter includes high-porosity contact-cemented sand. It is remarkable that a model introduced in 1980 is supported by at least two later datasets. On the other hand, RHG strongly overestimates the velocity measured in Han's (1986) unconsolidated Ottawa sand as well as the velocity measured by Blangy (1992) in high-porosity friable sand from the North Sea Troll field. In Figure 2.7 we compare the wet-rock RHG predictions with data and include Han's (1986) data in the entire range of clay content. These predictions are quite accurate for competent rock but, once again, strongly overestimate the friable-sand data.

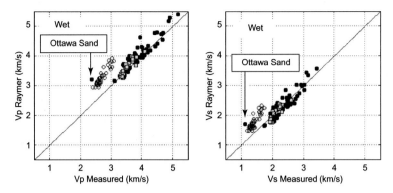

Figure 2.7 Velocity in wet rock as predicted by Eqs (2.23) and (2.24) versus the corresponding data (computed from the dry-rock data using Gassmann's fluid substitution). Left: The *P*-wave velocity. Right: The *S*-wave velocity. The symbols correspond to the same datasets as used in Figure 2.6. In this example, we use Han's (1986) data in the entire range of clay content, from zero to about 50%.

The RHG functional form was originally established for V_p. Much later, Dvorkin (2008a) showed that by applying the same functional form to V_s, accurate estimates of the shear-wave velocity are possible in multi-mineral rock, once again, excluding friable sand:

$$V_s = (1-\phi)^2 V_{ss} \sqrt{\frac{(1-\phi)\rho_s}{(1-\phi)\rho_s + \phi\rho_f}}, \tag{2.24}$$

where V_{ss} is the shear-wave velocity in the mineral phase; ρ_s is the density of the mineral phase; and ρ_f is the density of the pore fluid. V_{ss} can be estimated from Hill's average (Eq. (2.8)) while the densities can be estimated using the arithmetic average of the densities of the components as in Eq. (2.9).

Velocity predictions for wet-rock using Eq. (2.24) are compared to laboratory data in Figure 2.7 (right). As for V_p, we observe a fairly accurate match for competent sandstone and a strong overprediction of V_s in friable sand.

2.7   Other *S*-wave velocity predictors

For a known V_p, a number of empirical V_s predictors have been developed and can be used instead of Eq. (2.24). Moreover, they can be combined with the RHG V_p predictions to predict V_s not from V_p but from porosity, mineralogy, and pore-fluid properties. The V_s predictor equations in this section are applicable to wet (fully water-saturated rock) unless stated otherwise.

Pickett (1963) showed that in limestone $V_s = V_p / 1.9$, while in dolomite $V_s = V_p / 1.8$. Castagna *et al.* (1993) modified these relations as $V_s = -0.055 V_p^2 + 1.017 V_p - 1.031$ for limestone and $V_s = 0.583V_p - 0.078$ for dolomite, where the velocity is in km/s. In the same paper, the equation for clastic rock reads $V_s = 0.804V_p - 0.856$. The famous Castagna *et al.* (1985) "mudrock line" applicable to clastics gives $V_s = 0.862V_p - 1.172$.

Han (1986) used a large experimental dataset of sandstones with wide ranges of porosity and clay content variation to obtain $V_s = 0.794V_p - 0.787$. These measurements were conducted on wet rock at ultrasonic frequency. Mavko *et al.* (2009) added to these measurements a number of data points from high-porosity unconsolidated sands: $V_s = 0.79V_p - 0.79$. Further analysis of Han's (1986) data provides $V_s = 0.754V_p - 0.657$ for rock where the clay content is below 0.25 and $V_s = 0.842V_p - 1.099$ where it exceeds 0.25. If the same dataset is parted according to porosity, it gives $V_s = 0.853V_p - 1.137$ for porosity below 0.15 and $V_s = 0.756V_p - 0.662$ for porosity above 0.15.

Williams (1990) used well log data to arrive at $V_s = 0.846V_p - 1.088$ for water-bearing sands and $V_s = 0.784V_p - 0.893$ for shales.

Greenberg and Castagna (1992) combined relations for various lithologies to provide a unified empirical transform in multi-mineral brine-saturated rock composed of sandstone, limestone, dolomite, and shale. Their prediction can be also used for rock with any pore fluid, if we assume that the shear modulus is not influenced by the pore fluid and apply Gassmann's fluid substitution to the bulk modulus. This equation reads

$$V_s = \frac{1}{2}\left\{\left[\sum_{i=1}^{L} f_i \sum_{j}^{N_i} a_{ij} V_p^j\right] + \left[\sum_{i=1}^{L} f_i \left(\sum^{N_i} a_{ij} V_p^j\right)^{-1}\right]^{-1}\right\}, \quad \sum_{i=1}^{L} f_i = 1, \tag{2.25}$$

where L is the number of pure-mineral lithologic constituents; f_i are the volume fractions of these constituents in the whole mineral phase; a_{ij} are empirical coefficients; N_i is the order of polynomial for constituent i; V_p is the measured P-wave velocity; and V_s is the predicted S-wave velocity. The velocity is in km/s. The coefficients a_{ij} are given in Table 2.3. Vernik *et al.* (2002) modified this model to account for the V_p–V_s behavior in soft sediment (see discussion on Vernik's equations in Mavko *et al.*, 2009).

A different, theoretically motivated, approach to V_s prediction simply assumes that in dry rock the ratio of the bulk to shear modulus is exactly the same as in the solid (mineral) phase. This automatically means that $V_{sDry} / V_{pDry} = V_{ss} / V_{ps}$, where V_{pDry} and V_{sDry} are the velocities in the dry rock.

This assumption approximately complies with Pickett's (1963) equations and was first utilized by Krief *et al.* (1990). Expressions for saturated rock can be obtained simply by combining this relation with Gassmann's equations. Instead, Krief *et al.* (1990) suggested $(V_{pWet}^2 - V_{pf}^2) / V_{sWet}^2 = (V_{ps}^2 - V_{pf}^2) / V_{ss}^2$, where V_{pWet} and V_{sWet} are the P- and S-wave velocity in fully water-saturated rock, respectively.

The key purpose of the above V_s equations is to sensibly obtain V_s from V_p, the latter measured, for example, in the well, and then apply this quantification to

Table 2.3 *Regression coefficients for Eq. (2.21). These coefficients are only valid if the velocity is in km/s.*

Lithology	a_{i2}	a_{i1}	a_{i0}
Sandstone	0	0.80416	−0.85588
Limestone	−0.05508	1.01677	−1.03049
Dolomite	0	0.58321	−0.07775
Shale	0	0.76969	−0.86735

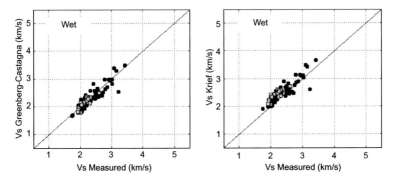

Figure 2.8 Velocity in wet rock as predicted by (left) the Greenberg and Castagna (1992) and (right) Krief *et al.* (1990) equations by combining these equations with RHG and calculating the velocity from porosity and clay content as explained in the text. The symbols are the same as used in Figure 2.7. The friable-sand data, including Ottawa sand, are not displayed.

forward modeling and/or interpreting AVO data. Notice that these equations can also be applied to model-predicted V_p (e.g., from RHG) and, hence, used as V_s predictors not from measured V_p but directly from porosity, mineralogy, and pore-fluid properties.

In Figure 2.8, we use this approach and compare laboratory data to V_s predicted from porosity and mineralogy (clay content) for wet competent rock, by first using RHG to compute V_p and then applying separately the Greenberg and Castagna (1992) and Krief *et al.* (1990) transforms to this V_p prediction. The former predictor appears to be more accurate than the latter.

2.8   Contact-cement model

RHG matches the Strandenes (1991) data (Figure 2.6) in high-porosity sand while it overestimates the Blangy (1992) data in the same porosity range between 0.25 and

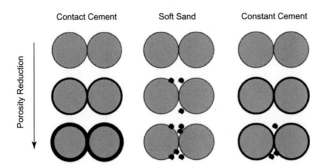

Figure 2.9 Three conceptual models of porosity reduction in a pack of spherical grains. From left to right: contact cement, soft (uncemented) sand, and constant cement. The arrow shows the direction of porosity reduction.

0.35. The mineralogy of these two datasets is very similar – they have very small amounts of clay and consist of stiffer minerals, predominantly quartz, mica, and feld-spar. The question is then why the velocity differs so much between these two sets. A plausible hypothesis is that this is due to the cemented grain contacts in the first set and the absence of such cementation in the second set. To advance this hypothesis, Dvorkin and Nur (1996) offered a micromechanical model of a granular sediment where the initial high porosity of a pack of identical spherical grains reduces due to diagenetic cement accumulating around the grains (Figure 2.9). The exact solution of this problem is an ordinary integro-differential equation which has to be solved to find the normal and tangential stiffnesses of a two-grain system. Once this is resolved, a simple statistical approach is used to calculate the elastic moduli of an aggregate of such grains, assuming that the local stress components at each grain are the same as given by the effective stress tensor in the entire composite. This is a strong assumption that can sometimes be not valid (Sain, 2010).

The equations describing the elastic behavior of this system (room-dry) and for this mean-field-approximation assumption are

$$K_{Cem} = \frac{1}{6}n(1-\phi_c)M_cS_n, \quad G_{Cem} = \frac{3}{5}K_{Cem} + \frac{3}{20}n(1-\phi_c)G_cS_\tau, \tag{2.26}$$

where K_{Cem} and G_{Cem} are the effective bulk and shear moduli of the cemented aggregate, respectively; n is the coordination number (the average contact number per grain) in the original high-porosity pack (about 6 or 8); M_c and G_c are the compressional and shear moduli of the cementing mineral (e.g., quartz, calcite, clay, or mixtures thereof), respectively; ϕ_c is the critical porosity (the porosity of the original uncemented grain pack); and S_n and S_τ are given by the following equations:

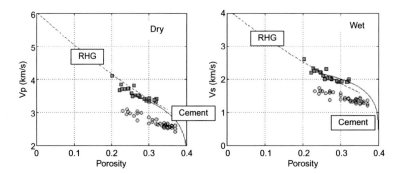

Figure 2.10 Velocity versus porosity in wet rock. The data used are the same as in Figure 2.6. The dashed curve is RHG. The solid curve is from the contact-cement model.

$$S_n = A_n(\Lambda_n)\alpha^2 + B_n(\Lambda_n)\alpha + C_n(\Lambda_n),$$

$$A_n(\Lambda_n) = -0.024153 \cdot \Lambda_n^{-1.3646}, \quad B_n(\Lambda_n) = 0.20405 \cdot \Lambda_n^{-0.89008},$$

$$C_n(\Lambda_n) = 0.00024649 \cdot \Lambda_n^{-1.9864};$$

$$S_\tau = A_\tau(\Lambda_\tau, v)\alpha^2 + B_\tau(\Lambda_\tau, v)\alpha + C_\tau(\Lambda_\tau, v), \tag{2.27}$$

$$A_\tau(\Lambda_\tau, v) = -10^{-2} \cdot (2.26v^2 + 2.07v + 2.3) \cdot \Lambda_\tau^{0.079v^2 + 0.1754v - 1.342},$$

$$B_\tau(\Lambda_\tau, v) = (0.0573v^2 + 0.0937v + 0.202) \cdot \Lambda_\tau^{0.0274v^2 + 0.0529v - 0.8765},$$

$$C_\tau(\Lambda_\tau, v) = 10^{-4} \cdot (9.654v^2 + 4.945v + 3.1) \cdot \Lambda_\tau^{0.01867v^2 + 0.4011v - 1.8186};$$

and

$$\Lambda_n = \frac{2G_c}{\pi G} \frac{(1-v)(1-v_c)}{1-2v_c}, \quad \Lambda_\tau = \frac{G_c}{\pi G}, \quad \alpha = [\frac{2(\phi_c - \phi)}{3(1-\phi_c)}]^{0.5}, \tag{2.28}$$

where G and v are the shear modulus and Poisson's ratio of the original-grain material, respectively, and G_c and v_c are those of the cementing mineral. These equations describe the best-fit approximation for the solutions of the integro-differential equations of the problem posed.

The contact-cement curve for quartz grains and quartz cement and $n = 6$ and $\phi_c = 0.4$ is shown in Figure 2.10. This model explains the rapid increase of velocity as the porosity decreases from ϕ_c.

2.9 Soft-sand model

The soft-sand model of Dvorkin and Nur (1996) is also called the modified lower Hashin–Shtrikman bound. This model intends to heuristically describe the elastic behavior of a pack of identical elastic spheres where porosity reduction is due to the introduction of non-cementing particles into the pore space (Figure 2.9, middle).

The soft-sand model connects two endpoints in the velocity–porosity plane: the high-porosity endpoint is at the critical porosity, ϕ_c while the zero-porosity endpoint is simply the velocity in the nonporous mineral matrix which can be a mixture of various pure mineralogical components.

The elastic moduli of the original room-dry grain pack at ϕ_c can be estimated from, for example, the Hertz–Mindlin contact theory (Mindlin, 1949) as

$$K_{HM} = \left[\frac{n^2(1-\phi_c)^2 G^2}{18\pi^2(1-v)^2} P \right]^{\frac{1}{3}}, \quad G_{HM} = \frac{5-4v}{5(2-v)} \left[\frac{3n^2(1-\phi_c)^2 G^2}{2\pi^2(1-v)^2} P \right]^{\frac{1}{3}}, \tag{2.29}$$

where P is the hydrostatic confining pressure applied to the pack and the other notations are the same as in Eqs (2.26) and (2.28). In Eq. (2.29) it is assumed that the grains have infinite friction force (no slip) at their contacts. If we allow only the fraction f of these contacts to have infinite friction while the rest of the contacts are frictionless and can slip, the equation for K_{HM} does not change but G_{HM} becomes now

$$G_{HM} = \frac{2+3f-v(1+3f)}{5(2-v)} \left[\frac{3n^2(1-\phi_c)^2 G^2}{2\pi^2(1-v)^2} P \right]^{\frac{1}{3}}. \tag{2.30}$$

We call parameter f the shear modulus correction factor.

Finally, at any porosity $\phi < \phi_c$

$$K_{Soft} = \left[\frac{\phi/\phi_c}{K_{HM}+\frac{4}{3}G_{HM}} + \frac{1-\phi/\phi_c}{K+\frac{4}{3}G_{HM}} \right]^{-1} - \frac{4}{3}G_{HM},$$

$$G_{Soft} = \left[\frac{\phi/\phi_c}{G_{HM}+z_{HM}} + \frac{1-\phi/\phi_c}{G+z_{HM}} \right]^{-1} - z_{HM}, \quad z_{HM} = \frac{G_{HM}}{6} \left(\frac{9K_{HM}+8G_{HM}}{K_{HM}+2G_{HM}} \right). \tag{2.31}$$

Notice that the high-porosity endpoint in this model does not necessarily have to be governed by Eqs (2.29) and (2.30). It can simply be selected from, for example, relevant experimental data obtained either in the laboratory or a well. The main point of this model is the "soft" connector given by Eq. (2.31).

The soft-sand curves for V_p and V_s and for pure quartz grains at 30 MPa confining pressure are displayed in Figure 2.11 (for $n = 7$ and $\phi_c = 0.4$). These curves are computed for wet rock by first using the soft-sand model for room-dry grain pack and then applying Gassmann's fluid substitution to these results.

2.10   Stiff-sand model

The two endpoints discussed in the soft-sand model can also be connected by a "stiff" connector, also called the modified upper Hashin–Shtrikman bound or the stiff-sand model. The appropriate equation (Gal et al., 1998) is

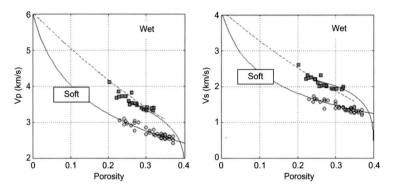

Figure 2.11 Same as Figure 2.10 but with the soft-sand curve added.

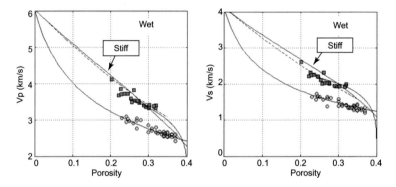

Figure 2.12 Same as Figure 2.11 but with the stiff-sand curve added.

$$K_{Stiff} = \left[\frac{\phi/\phi_c}{K_{HM} + \frac{4}{3}G} + \frac{1 - \phi/\phi_c}{K + \frac{4}{3}G} \right]^{-1} - \frac{4}{3}G,$$

$$G_{Stiff} = \left[\frac{\phi/\phi_c}{G_{HM} + z} + \frac{1 - \phi/\phi_c}{G + z} \right]^{-1} - z, \quad z = \frac{G}{6}\left(\frac{9K + 8G}{K + 2G} \right). \tag{2.32}$$

The stiff-sand curves for V_p and V_s and for pure quartz grains at 30 MPa confining pressure are displayed in Figure 2.12. They are computed in the same fashion as the soft-sand curves displayed in Figure 2.11.

It is remarkable that these theoretical curves are practically the same as predicted by the much earlier RHG model calibrated to competent stiff rock.

2.11   Constant-cement model

Finally, imagine that a high-porosity grain pack has some initial cementation but any further porosity reduction is due to the introduction of non-cementing material into the

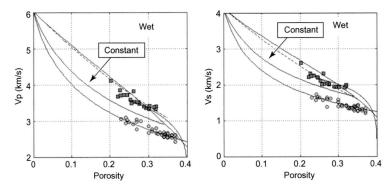

Figure 2.13 Same as Figure 2.12 but with constant-cement curves added.

pore space (Figure 2.9, right). The high-porosity endpoint can be selected on the contact-cement curve, RHG, or stiff-sand curve and then connected to the zero-porosity endpoint by the soft connector according to Eqs (2.27). The resulting formulae are

$$
K_{Const} = \left[\frac{\phi / \phi_c}{K_{Cem} + \frac{4}{3} G_{Cem}} + \frac{1 - \phi / \phi_c}{K + \frac{4}{3} G_{Cem}} \right]^{-1} - \frac{4}{3} G_{Cem},
$$

$$
G_{Const} = \left[\frac{\phi / \phi_c}{G_{Cem} + z_{Cem}} + \frac{1 - \phi / \phi_c}{G + z_{Cem}} \right]^{-1} - z_{Cem}, \quad z_{Cem} = \frac{G_{Cem}}{6} \left(\frac{9K_{Cem} + 8G_{Cem}}{K_{Cem} + 2G_{Cem}} \right),
$$

(2.33)

where K_{Cem} and G_{Cem} are the bulk and shear moduli, respectively, of the original slightly cemented pack.

An easy way of systematically obtaining K_{Cem} and G_{Cem} is simply to use the Hertz–Mindlin equations and assume an unrealistically high coordination number n (e.g., $n = 15$ or 21). This is purely mathematical convenience, since a pack of identical spherical grains cannot have a coordination number this high. All it allows us to accomplish is to change the high-porosity endpoint to make it start on the contact cement or stiff-sand model curve. The constant-cement model curve thus obtained should not be extended into the porosity domain above its intersection with the stiff-sand or contact-cement curve. Such constant-cement curves computed using the soft-sand model but with $n = 15$ instead of $n = 6$ are displayed in Figure 2.13. Avseth *et al.* (2000) show that such velocity–porosity curves can accurately describe the elastic behavior of real sediment deposited in a turbidite channel.

2.12   Inclusion models

All micromechanical models described here assume that the rock is formed by solid grains which comprise an uncemented grain pack at the critical porosity and, as the porosity decreases, the original grain pack is altered either by grain-contact cement (the diagenetic trend) or by smaller grains deposited in the pore space between the original larger grains (the sorting trend). It is quite surprising that such highly idealized representations of real rock produce the elastic moduli which sometimes match laboratory and field data. This fact further supports the sentiment expressed by statistician George Box, although in a different context, that "all models are wrong, but some are useful" (Box and Draper, 1987).

Bearing this situation in mind, let us introduce a different class of micromechanical models, the inclusion models. A detailed review of these models that build a rock from the zero-porosity endpoint (rather than from the critical porosity) by placing inclusions into the solid matrix can be found in Mavko *et al.* (2009). These models are perhaps relevant to some carbonate rocks where the pores appear as inclusions in a calcite or dolomite matrix.

We illustrate this class of models by the differential effective-medium model (DEM). DEM does not have a closed-form solution. Rather, a coupled system of ordinary differential equations has to be solved to obtain the effective elastic moduli of a solid with inclusions (see Mavko *et al.*, 2009).

DEM requires four inputs: the elastic moduli of the rock's mineral matrix; the elastic moduli of the inclusions; porosity; and the aspect ratio of the inclusions, the ratio of the short to long axis where it is assumed that the two long axes of an inclusion are equal. If the inclusions are filled with fluid, only its bulk modulus is required. However, direct DEM computation of the elastic moduli of rock with fluid will provide results relevant to high-frequency velocity measurements (such as using the ultrasonic pulse technique in the laboratory). The reason is that DEM treats the inclusions as disconnected from each other. Hence, as a model porous solid is excited by a propagating elastic wave, the resulting small pressure increments in the pores cannot equilibrate within a finite rock volume as happens in real rock at low-frequency excitation. A mitigating recipe is simple: first compute the effective elastic moduli of a completely dry porous solid and then conduct fluid substitution using Gassmann's equation.

Figure 2.14 shows DEM curves computed for wet pure-quartz porous rock and progressively decreasing aspect ratio. One difference between the display in this figure and other velocity–porosity cross-plots in this section is that here the porosity range extends from zero to 100% rather than from zero to the critical porosity, which is about 0.40 for clastic sediment. Extending the porosity range is reasonable for carbonates, as chalk porosity may reach 50% (e.g., Fabricius *et al.*, 2002). The user of this model has to be aware that the solution may degenerate at high porosity, especially for a small aspect ratio.

Still, the example in Figure 2.14 indicates that DEM provides realistic velocity values for an aspect ratio about 0.10: this model curve is close to the RHG–Dvorkin predictions (Eqs (2.23) and (2.24)) for the same mineralogy. In the V_p–porosity graph, the RHG prediction is exactly matched by the DEM curve with 0.13 aspect ratio. The

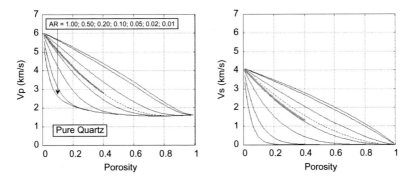

Figure 2.14 Elastic-wave velocity DEM model curves (solid black) for pure quartz with inclusions filled with water (the bulk modulus and density of water are listed in Table 2.2) and gradually decreasing aspect ratio (AR), as marked in the plot. The computations were conducted for dry rock following Gassmann's fluid substitution. Bold gray curves are from RHG–Dvorkin model. Dashed black curves are from DEM with aspect ratio 0.13.

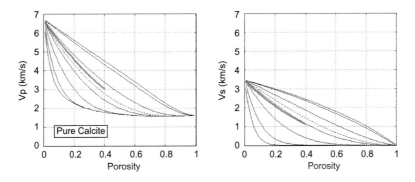

Figure 2.15 Same as Figure 2.14 but for pure calcite.

same aspect ratio slightly overpredicts the RHG–Dvorkin curve in the V_s–porosity graph.

An example for pure-calcite mineralogy is shown in Figure 2.15. In this case, the RHG model curve cannot be exactly matched by the DEM model with aspect ratio 0.13. This manifests the limitations in the correspondence between different rock physics models (see exhaustive analysis of this correspondence in Ruiz, 2009).

Finally, to once again show the possibilities and limitations of correspondence between different models, in Figure 2.16, we duplicate Figure 2.15 but instead of plotting the RHG–Dvorkin model curve, we plot the soft- and stiff-sand model curves computed for pure calcite with water, differential pressure 30 MPa, coordination number 6, critical porosity 0.40, and the shear modulus correction factor $f = 1$.

Inclusion models appear to be natural for carbonates and other lithologies, such as volcanic rocks, where the pore space looks like inclusions rather than the space between distinctive grains. However, this perception should not prevent us from applying the

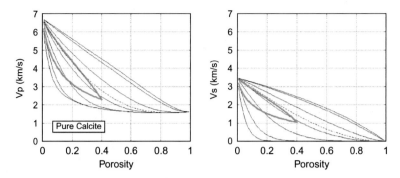

Figure 2.16 Same as Figure 2.15 but with the soft- and stiff-sand curves (bold gray), both for differential pressure 30 MPa, coordination number 6, critical porosity 0.40, and the shear modulus correction factor 1.

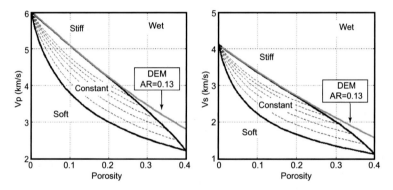

Figure 2.17 Same as Figure 2.13 but without data and with the soft, stiff, and constant-cement curves displayed. The former two curves (displayed as solid) are computed for the coordination number 6. The latter curves (dashed) are computed for gradually increasing coordination number, from 10 to 30 with increment 5 to simulate the increasing degree of the initial cementation. Bold gray curve is from DEM with aspect ratio 0.13.

granular-rock models to carbonates because (a) in some carbonates, distinctive granular structure is apparent and (b) different types of models can be used to arrive at approximately the same elastic properties.

2.13 Summary of the models

RHG, soft-sand, stiff-sand, and constant-cement models will be used in this book to produce the elastic properties of sediment from the primary properties, porosity, mineralogy, texture, and pore fluid. To further illustrate the connection between the latter three models, we compute and display the relevant curves in Figure 2.17. Figure 2.18 graphically illustrates the meaning of these models.

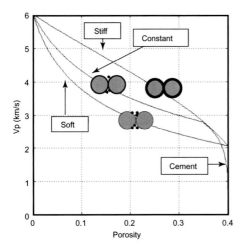

Figure 2.18 A scheme of velocity–porosity behavior for different scenarios of porosity reduction.

2.14 Properties of the pore fluid phases

The two inputs required to determine the bulk modulus and density of the pore fluid that is an immiscible mixture of two or more fluid phases are the bulk moduli and densities of these phases. Batzle and Wang (1992) provide empirical equations that relate these properties of brine, oil, and gas to the brine's salinity; oil's API gravity; gas-to-oil ratio (GOR); gas gravity; pore pressure and temperature (see also Mavko *et al.*, 2009).

Examples of such computed properties are shown in Figure 2.19.

2.15 A note on effective and total porosity and fluid substitution*

The velocity–porosity effective-medium and empirical models presented here operate with the total porosity, ϕ, which is defined as the ratio of the total void volume to the total bulk volume of a sample. At the same time, the effective porosity, ϕ_e, is often used in petrophysical and petroleum engineering applications. ϕ_e has several definitions, one of which is the ratio of the void space of porous rock capable of transmitting a fluid to the total volume. Porosity that is not considered effective porosity includes water bound to clay (shale) particles as well as isolated (vuggy) porosity in carbonates or low-porosity sandstone.

Using the effective porosity can be appropriate for fluid substitution in shaly sand where the shale part of the rock is always wet and has extremely low permeability, so

---

* This part was modified from work originally published by SEG (Dvorkin *et al.*, 2007)

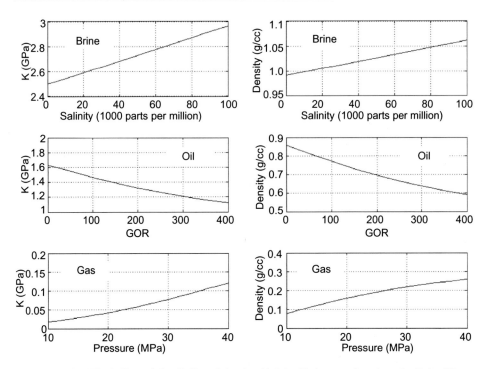

Figure 2.19 Top: The bulk modulus (left) and density (right) of brine as a function of salinity. The pore pressure is 20 MPa and temperature is 60° C. Middle: The bulk modulus (left) and density (right) of oil as a function of GOR. The oil's API gravity is 30, the pore pressure is 20 MPa and temperature is 60° C. Bottom: The bulk modulus (left) and density (right) of gas as a function of pore pressure. Gas gravity is 0.65 and temperature is 60° C.

that the pore fluid can only be replaced in the effective-porosity part of the pore space. Moreover, conducting fluid substitution in the total pore space of shaly rock may violate the main assumption of Gassmann's theory that the minuscule pore pressure fluctuation induced in the fluid by a seismic wave can rapidly equilibrate through the pore space: shale contains bound water, which is essentially immobile and, hence, cannot be in hydraulic equilibrium with the rest of the pore space.

Bearing this complication in mind, Dvorkin *et al.* (2007) offer a method for fluid substitution only in the effective-porosity part of the pore space. This method involves a number of assumptions and additional inputs. Still, it should be part of the arsenal of the rock physicist.

Consider a porous rock with clay (Figure 2.20). The volume fraction of the clay mineral in the whole mineral phase is f_{clay}. The intrinsic porosity of clay (the micro-porosity) is ϕ_{clay}. In a unit volume of rock, the volume occupied by the non-clay minerals (assumed to be nonporous) is $(1 - f_{clay})(1 - \phi)$, where, as before, ϕ is the total porosity. The volume occupied by the clay mineral is $f_{clay}(1 - \phi)$. Then the volume occupied by the porous clay is

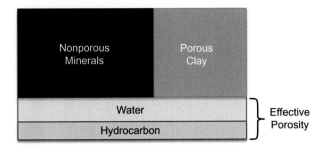

Figure 2.20 Schematic representation of the components of rock with porous clay, water, and hydrocarbon.

$$c = \frac{f_{clay}(1-\phi)}{1-\phi_{clay}}.$$ (2.34)

The total void space in the rock is the sum of the void space outside the porous clay and that within the clay. The void space outside the porous clay (in a unit volume of rock) is defined here as the effective porosity:

$$\phi_e = \phi - \phi_{clay}c = \phi - f_{clay}\phi_{clay}\frac{1-\phi}{1-\phi_{clay}}.$$ (2.35)

Let us introduce a modified solid phase that includes the non-clay minerals plus the porous clay. The volume of this modified solid phase in a unit volume of rock is $1-\phi_e$. The volume fraction of the porous clay in the modified solid is

$$f_{Pclay} = \frac{f_{clay}}{1-\phi_{clay}(1-f_{clay})}.$$ (2.36)

The part of the total pore space saturated with hydrocarbon is $S_h = 1 - S_w$. If the intrinsic pore space of the clay is fully water-saturated and all hydrocarbon is contained in the effective pore space, the volume fraction of the hydrocarbon in the effective-porosity pore space is

$$S_{he} = \phi(1-S_w)/\phi_e,$$ (2.37)

while the water saturation in the effective-porosity pore space is

$$S_{we} = 1 - S_{he} = 1 - \phi(1-S_w)/\phi_e.$$ (2.38)

As expected, S_{we} becomes zero if the only water in the rock is that contained within the clay, that is, $S_w = c\phi_{clay}/\phi$.

Fluid substitution equations have the same form as Eqs (2.12) and (2.13), but now we have to replace ϕ with ϕ_e as well as use the bulk moduli for the modified solid phase that includes the nonporous mineral plus porous clay and for the fluid that contains S_{we} volume fraction of water and S_{he} fraction of the hydrocarbons. Specifically, if the bulk modulus of the 100% wet rock, K_{Wet}, is known, the dry-rock modulus, K_{Drye}, of the rock where the effective-porosity pore space is empty but the clay is still 100% wet is

$$K_{Drye} = K_{se} \frac{1-(1-\phi_e)K_{Wet}/K_{se} - \phi K_{Wet}/K_w}{1+\phi_e - \phi_e K_{se}/K_w - K_{Wet}/K_{se}}, \tag{2.39}$$

where K_{se} is the bulk modulus of the modified solid phase and K_w is the bulk modulus of water. Conversely, the bulk modulus of rock with hydrocarbons can be computed from K_{Drye} as

$$K_{Sat} = K_{se} \frac{\phi_e K_{Drye} - (1+\phi_e)K_{fe}K_{Drye}/K_{se} + K_{fe}}{(1-\phi_e)K_{fe} + \phi_e K_{se} - K_{fe}K_{Drye}/K_{se}}, \tag{2.40}$$

where K_{fe}, the effective bulk modulus of the immiscible combination of water and hydrocarbon in the effective-porosity pore space, is

$$K_{fe} = [S_{we}/K_w + (1-S_{we})/K_h]^{-1}, \tag{2.41}$$

where K_h is the bulk modulus of the hydrocarbon.

The bulk modulus of the modified solid phase, K_{se}, can be estimated from Hill's average (Eqs (2.5) and (2.6)) for a two-component modified solid where one component is the nonporous mineral and the second one is the wet porous clay.

The volume fraction of the porous clay in the modified solid is f_{Pclay} (Eq. (2.36)) while that of the nonporous solid is $1 - f_{Pclay}$. Let us assume that the nonporous mineral is quartz with the bulk modulus K_q. Then the bulk modulus of the modified solid is

$$K_{se} = \frac{K_{sV} + K_{sR}}{2},$$
$$K_{sV} = f_{Pclay}K_{Pclay} + (1-f_{Pclay})K_q, \tag{2.42}$$
$$K_{sR} = [f_{Pclay}/K_{Pclay} + (1-f_{Pclay})/K_q]^{-1},$$

where K_{Pclay} is the bulk modulus of wet porous clay and is the only remaining parameter needed for our effective-porosity fluid substitution.

There are several ways of assessing this parameter. One is to simply locate a pure clay (shale) interval in a well and compute K_{Pclay} from the measured V_p, V_s, and ρ_b as $\rho_b(V_p^2 - \frac{4}{3}V_s^2)$. Another is to use one of rock physics models, for example the soft-sand model, and apply it only to the wet porous clay.

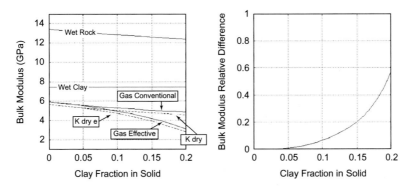

Figure 2.21 Left: Bulk modulus versus clay content in the solid phase for wet rock; wet porous clay; gas-saturated rock where fluid substitution was conducted with the effective porosity; and gas rock using conventional fluid substitution. Right: The difference between the bulk modulus of rock with gas with the effective porosity and that using conventional fluid substitution, normalized by the former. The total porosity is 0.20. The dashed curves in the left-hand plot are for the dry-rock bulk modulus computed using the method described here and conventional fluid substitution. The wet-rock bulk modulus was computed using the soft-sand model and the mixed clay/quartz mineralogy with the clay fraction in the solid phase varying from zero to 0.20. The effective pressure used in this model was 30 MPa, critical porosity 0.40, coordination number 6, and the shear-modulus correction factor 1. The wet-clay bulk modulus was computed using the same model but for pure-clay mineralogy.

In the following example, consider rock with the total porosity $\phi = 0.20$ and $f_{clay} = 0.20$. The rest of the mineral phase is pure quartz. The intrinsic porosity of clay $\phi_{clay} = 0.30$. Assume the wet-rock bulk modulus $K_{Wet} = 12.38$ GPa and water saturation $S_w = 0.30$ with the rest of the pore space occupied by gas. The fluid-phase properties are those from Table 2.2 while the mineral properties are from Table 2.1.

Under these circumstances, the effective porosity $\phi_e = 0.13$ and effective water saturation $S_{we} = 0.087$. The dry-rock effective bulk modulus, K_{Drye}, computed from Eq. (2.39) is 2.81 GPa and the bulk modulus of the same rock with gas is, according to Eq. (2.40), $K_{Sat} = 3.08$ GPa.

If we use *conventional fluid substitution* in this quartz/clay rock, the resulting bulk modulus at partial saturation is $K_{Sat} = 4.84$ GPa. The difference is large and may be important in assessing the fluid effect on seismic reflections. Of course, as the clay content, f_{clay}, reduces from 0.20 to zero, so does the difference between the bulk moduli computed according to these two methods (Figure 2.21).

This difference results from the difference in the dry-rock bulk modulus as computed from the given wet-rock modulus using the method described here (K_{Drye}) and conventional fluid substitution (K_{Dry} in Eq. (2.13)). These two moduli are also plotted in Figure 2.21. The results may change as a function of several factors. Figure 2.22 shows the results for the total porosity 0.30 and 0.40 keeping all the other inputs the same (as used in Figure 2.21).

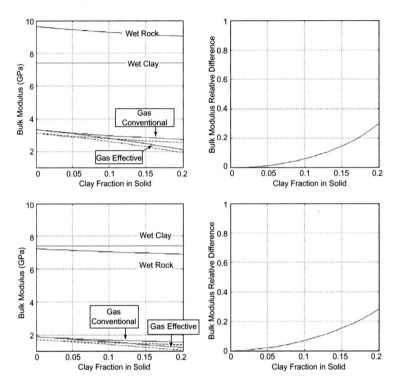

Figure 2.22 Same as Figure 2.21 but for the total porosity 0.30 (top) and 0.40 (bottom). The scale of the graphs is changed for clarity.

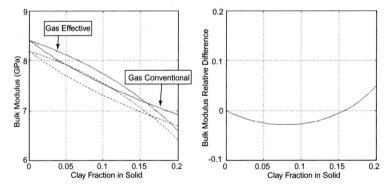

Figure 2.23 Same as Figure 2.21 but for stiffer rock (the wet-rock bulk modulus is about 14 GPa based on the constant-cement model using the soft-sand model with coordination number 12). All other parameters are the same as used for Figure 2.21. The scale of the graphs is changed for clarity.

Figure 2.23 shows the effect of stiffening the rock. In this example we used the constant-cement model (same as the soft-sand model but with coordination number 12) to obtain the bulk modulus of wet rock which, depending on the clay content varied from 14.90 (zero clay) to 13.50 GPa (20% clay). The difference between the two fluid

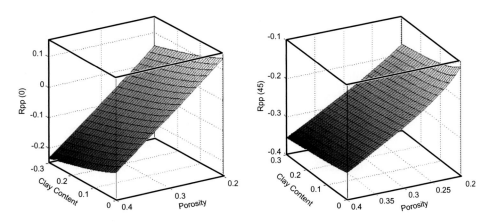

Figure 2.24 The reflection amplitude at the shale/sand interface at zero angle of incidence (left) and 45° angle of incidence (right) computed as explained in the text.

substitution methods is much smaller here as compared to our first example. Moreover, for the clay content between zero and about 15%, the rock-with-gas bulk modulus obtained using conventional fluid substitution is smaller than that obtained using the method discussed here. Still, at 20% clay content, the behavior is the same as observed in the previous examples: the rock-with-gas bulk modulus obtained using conventional fluid substitution is larger than that obtained using the method discussed here.

The inputs for this fluid substitution method have to be selected with caution to avoid negative effective porosity or effective water saturation as well as negative dry-rock bulk modulus.

2.16 Example of applying rock physics models to simulate seismic amplitude

To illustrate a link between the rock physics transforms and synthetic seismic amplitude, consider an interface between soft shale and partially cemented gas sand. Assume the porosity of the shale is 0.30 and its clay content is 0.60. In the sand, the porosity varies between 0.20 and 0.40 while the clay content is between zero and 0.30. The rock physics model for the shale is the soft-sand model with the coordination number 6 and the shear modulus correction factor 1. For the sand we choose the constant-cement model whose functional form is the same as the soft-sand model but with the coordination number 10 and shear modulus correction factor 1. The differential pressure for both shale and sand is 30 MPa. The critical porosity for both shale and sand is 0.40.

To compute the bulk moduli and densities of water and gas we select the salinity 70,000 ppm, gas gravity 0.70, pore pressure 20 MPa and temperature 80° C. According to the Batzle–Wang (1992) equations, the resulting bulk moduli and densities are, respectively, 2.80 GPa and 1.03 g/cm^3 for water and 0.04 GPa and 0.16 g/cm^3 for gas. Assume also that water saturation in the sand is 30%. The resulting bulk modulus and density for the gas/water fluid in the sand are 0.06 GPa and 0.421 g/cm^3, respectively.

The resulting reflection amplitudes at the shale/sand interface computed according to the Zoeppritz (1919) equations (see Chapter 4) at the incidence angles zero and 45° are plotted versus the sand's porosity and clay content in Figure 2.24.

The normal reflectivity for the lowest-porosity cleanest sand is positive. As the angle increases, it changes the phase and becomes negative. The highest-porosity sand shows negative normal reflectivity and increasingly negative reflectivity with the increasing angle of incidence.

This simple example shows an immediate use of rock physics modeling for AVO response prediction and real-data reconnaissance.

3 Rock physics diagnostics

3.1 Quantitative diagnostics

The process of finding a model that accurately mimics the site-specific elastic behavior of sediment is called rock physics diagnostics. It is performed on well log data and includes two steps: (a) bringing the entire interval under examination to a common fluid denominator by theoretically substituting the in situ pore fluid with, for example, the formation brine and (b) finding a model curve that fits these "wet" data.

The first example of rock physics diagnostics is given in Figure 3.1, where we display depth curves in a well examined in Dvorkin *et al.* (2004). This well has an oil-filled sandstone interval with porosity of about 0.20. The rest of the interval is shale. V_p calculated for wet conditions, as expected, exceeds the measured velocity in the sand while V_s remains practically unchanged.

By making velocity–porosity cross-plots (Figure 3.2) we find that, unlike in the examples discussed in Chapter 2, both V_p and V_s hardly increase with decreasing porosity. The reason is the simultaneous increase in the clay content and reduction in porosity. The first factor makes the rock softer while the second factor makes it stiffer. By interacting, these two factors produce almost flat velocity–porosity plots.

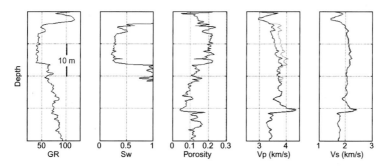

Figure 3.1 Depth curves in an oil well. From left to right: gamma-ray (GR); water saturation; total porosity; *P*-wave velocity as measured in situ (black) and theoretically computed by Gassmann's fluid substitution for wet conditions (gray); and the *S*-wave velocity in situ (black) and for wet conditions (gray).

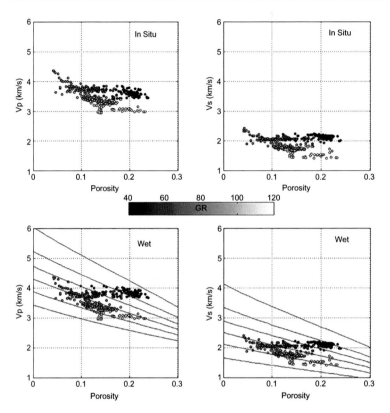

Figure 3.2 Velocity–porosity cross-plots of well data displayed in Figure 3.1 for in situ conditions (top) and wet conditions (bottom). The symbols are color-coded by GR (the lighter the symbol the higher the GR). GR is used here as a proxy for the clay content. The model curves are from the stiff-sand model as described in the text.

To find a model that explains these data, we first fluid-substitute the entire interval for wet conditions. The model curves, of course, will also be computed for wet conditions and with the same common fluid denominator (brine) as used in fluid substitution in the well.

The model selected in this case is the stiff-sand model with the coordination number 6, differential pressure 30 MPa, and parameter $f = 1$. Six model curves are superimposed on the wet-rock data in Figure 3.2. All of them are computed using the same model but for changing mineralogy: the top curve is for 100% quartz while the bottom curve is for 100% clay. The intermediate curves are computed for clay content varying from zero to 100% with a 20% increment (from top to bottom).

We observe that the model selected here quantitatively explains the P-wave data: the sand has the clay content between zero and 20%. The transition to shale is accompanied by the increasing clay content. As a result, part of the velocity–porosity trend appears flat. Now that an appropriate model has been established, it can be used to theoretically

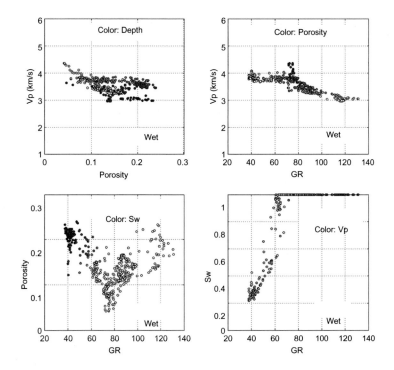

Figure 3.3 Data from Figure 3.1. The *P*-wave velocity corresponds to wet conditions. Clockwise from top-left: velocity versus porosity color-coded by depth (dark is for smaller depth and light is for larger depth); velocity versus GR color-coded by porosity (dark is for smaller porosity and light is for larger porosity); water saturation versus GR color-coded by the velocity (dark is for lower velocity and light is for higher velocity); and porosity versus GR color-coded by water saturation (dark is partial saturation and light is full saturation).

alter porosity, mineralogy, and, of course, the pore fluid. By doing this alteration in a geologically consistent fashion, we can create pseudo-wells where, for example, the sand interval tapers off and is gradually replaced by shale. Synthetic seismograms computed at such a pseudo-well (similar to the catalogue shown in Figure 1.9) can aid in interpreting real seismic data away from well control.

The cross-plots shown in Figure 3.2 are in essence three-dimensional plots, where the third dimension is the color (GR in this example). Other parameters can be used to better understand the depositional processes that created an interval under examination. As an example, consider Figure 3.3 where we use the computed wet-rock velocity and cross-plot (a) V_p versus porosity, color-coded by depth; (b) V_p versus GR, color-coded by porosity; (c) water saturation versus porosity, color-coded by V_p; and (d) porosity versus GR, color-coded by water saturation.

From the first cross-plot we observe that the soft high-porosity shale is located at the top of the interval while the shale located below the sand undergoes gradual

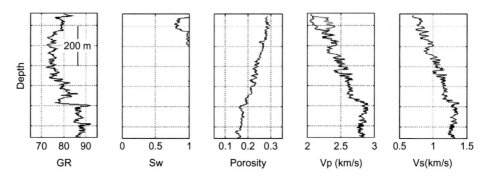

Figure 3.4 Same as Figure 3.1 but for a different well. The curves have been smoothed.

compaction. The second cross-plot indicates that the velocity generally decreases with increasing GR except for a low-porosity high-velocity spike in the lower part of the well. The third cross-plot shows, as expected, that hydrocarbons are present only in the low-GR interval. The fourth and final cross-plot is especially important as it shows us a complex porosity–GR behavior. This trend is *V*-shaped and illustrates a gradual transition from medium-porosity relatively clean sand to shaly sand accompanied by porosity reduction. After reaching the lowest-porosity point (at a GR of about 80), the porosity starts to increase with increasing GR until it reaches the relatively high-porosity pure shale point. This trend may help us vary the porosity and clay content in a coherent fashion consistent with this depositional environment.

Our second rock physics diagnostics example is for a well with small gas saturation at the top (Figure 3.4). The diagnostics cross-plots in Figure 3.5 indicate that the data can be described by the soft-sand model curves for quartz/clay mineralogy calculated for the coordination number 6, differential pressure 30 MPa, and $f = 2$. Strictly speaking, the differential pressure input should vary with depth; however, for the purpose of finding the appropriate model, we need to keep some of the parameters fixed. This is why in this example we use a constant differential pressure. This approach is, of course, justified where we examine relatively short depth intervals but has to be used with caution where we attempt to describe rock properties in a large depth interval by a single model without adjusting the inputs.

Figure 3.6 illustrates the importance of bringing the entire interval to a common fluid denominator: the velocity–porosity trend in the gas sand lines up with the trend in the shale at in situ conditions but if calculated for wet-rock conditions it lines up with the trend for the other (wet) sand interval.

The rock physics diagnostics procedure described here will be used in this book to establish rock physics transforms, create pseudo-wells, and then produce synthetic seismic catalogues. This procedure is based on trials and is possible only because we have a number of rock physics models in our arsenal. To avoid obtaining a good fit for the wrong reasons, it is very important to bear in mind the meaning of a model in terms of geology, deposition, and diagenesis.

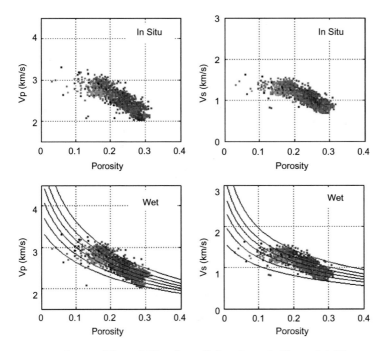

Figure 3.5 Same as Figure 3.2 but for well data shown in Figure 3.4. The curves are from the soft-sand model with the clay content increasing from the top to bottom curve from zero to 100% with a 20% increment. Color-code is GR.

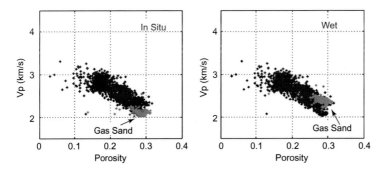

Figure 3.6 Importance of common fluid denominator. Velocity versus porosity for data displayed in Figure 3.3. Left: in situ conditions. Right: wet-rock conditions. The gray symbols are for the gas sand.

3.2   Qualitative diagnostics: staring at the data

Plotting data in various coordinates and color-codes and contemplating the results helps understand the underlying relations between the variables, such as the elastic properties, porosity, mineralogy, depth, and compaction. The following graphs (Figures

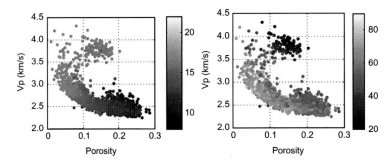

Figure 3.7 Velocity versus porosity color-coded by depth (left) and GR (right). The depth is in kft. The plot on the left shows compaction in the shale. The sand (dark in the plot on the right) is located deep in the interval and is well-cemented.

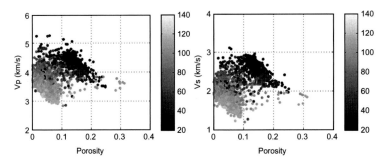

Figure 3.8 Velocity versus porosity color-coded by GR and showing low-porosity compacted shale and medium-porosity well-cemented sand.

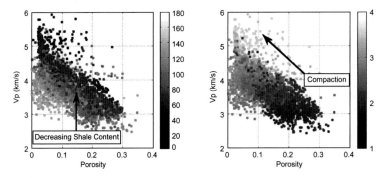

Figure 3.9 Velocity versus porosity color-coded by GR (left) and depth in km (right). The outlying data points in these cross-plots may indicate lower-quality data, but they should not be excluded from the analysis unless we understand the reason (e.g., distorted borehole geometry, which can be inferred from a caliper log).

3.7, 3.8, and 3.9) based on well measurements illustrate this idea. The figure captions explain the relations implied by the data.

3.3   Word of caution when using well data

Often, rock physics analysis of well data starts after the data were analyzed and processed by petrophysicists. The data thus altered may contain undesired artifacts. For example, the saturation curve may be directly derived from resistivity without regard to the fact that sometimes high resistivity may be due to, for example, small porosity. This may lead to hydrocarbon saturation larger than zero in shale that should be 100% wet. The S-wave velocity curve may be altered (or simply created from scratch) using one of V_s predictors. The total porosity may be derived from V_p by using, for example, Wylie's time average or Raymer's transform (the so-called sonic porosity). The rock physicist should always be aware of how a specific curve has been derived and conduct rock physics diagnostics accordingly, relying on the recorded rather than derived data.

Part II

Synthetic seismic amplitude

4 Modeling at an interface: quick-look approach

4.1 Reflection modeling at an interface: the concept

Forward modeling of seismic reflections at an interface between two elastic half-spaces is a traditional way of setting expectations for the character of seismic traces between the overburden shale and sand reservoir; at gas/oil, gas/water, and oil/water contacts; as well as at various unconformities present in the subsurface. To conduct such computations, the elastic properties of both half-spaces are required. If we know the site-specific transforms between the rock properties and conditions and the elastic properties, we can compute seismic reflections at an interface as a function of porosity, lithology, and fluid. In the next section of this chapter we will review the mathematical apparatus used to compute seismic reflections and then proceed with utilizing these equations for assessing the seismic signatures from the properties of the rock half-spaces forming the interface.

4.2 Normal reflectivity and reflectivity at an angle

The reflectivity at an interface between two elastic bodies is defined as the ratio of the reflected wave amplitude to the incident wave amplitude. As the wave strikes the interface, it produces the reflected and transmitted waves (Figure 4.1). Here we will analyze only the *reflected* P-wave. The incident P-wave can approach the interface in the direction normal to the interface or at a non-zero angle (Figure 4.1). The angle of incidence is defined as the angle between the direction of propagation of the wave front and the direction normal to the interface between the two half-spaces. While a normal-incidence P-wave does not produce S-waves, a P-wave at a non-zero incident angle produces reflected and transmitted S-waves. In the following equations for the P-to-P reflectivity, the properties of the upper interface are marked by subscript "1" while those of the lower interface are marked by subscript "2."

The equation for the amplitude of the reflected P-wave is especially simple for normal incidence (Zoeppritz, 1919):

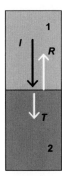 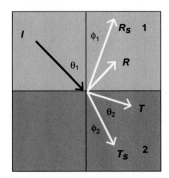

Figure 4.1 Left: *P*-wave normal to the interface (*I*) and the reflected (*R*) and transmitted (*T*) *P*-waves. Right: *P*-wave at a non-zero angle of incidence. θ_1 is the angle of incidence while θ_2 is the angle of the transmitted wave. The upper elastic half-space is marked by number "1" while the lower half-space is marked by number "2." The reflected and transmitted *S*-waves are marked "R_s" and "T_s," respectively. The angles of the reflected and transmitted *S*-waves are ϕ_1 and ϕ_2, respectively.

$$R_{pp}(0) = \frac{\rho_2 V_{p2} - \rho_1 V_{p1}}{\rho_2 V_{p2} + \rho_1 V_{p1}} = \frac{I_{p2} - I_{p1}}{I_{p2} + I_{p1}} \approx \frac{1}{2} \ln \frac{I_{p2}}{I_{p1}}, \tag{4.1}$$

where $R_{pp}(0)$ is the normal-incidence *P*-to-*P* reflectivity (zero angle of incidence). Equation (4.1) indicates that the normal reflectivity only depends on the impedances of the half-spaces.

The angle of the reflected *P*-wave is the same as the angle θ_1 of the incident *P*-wave. The angle θ_2 of the transmitted *P*-wave is determined from the following equation:

$$\sin \theta_2 = \sin \theta_1 \frac{V_{p2}}{V_{p1}}. \tag{4.2}$$

The equation for the amplitude of the reflected *P*-wave at a non-zero angle of incidence θ_1 is fairly complicated (Zoeppritz, 1919; Aki and Richards, 1980):

$$R_{pp}(\theta_1) = \left[\left(b \frac{\cos \theta_1}{V_{p1}} - c \frac{\cos \theta_2}{V_{p2}} \right) F - \left(a + d \frac{\cos \theta_1}{V_{p1}} \frac{\cos \phi_2}{V_{s2}} \right) H p^2 \right] / D, \tag{4.3}$$

where angle ϕ_2 is that of the transmitted *S*-wave (Figure 4.1). This angle, ϕ_2, as well as the angle ϕ_1 of the reflected *S*-wave and the ray parameter, *p*, are determined from the following equation:

$$p = \frac{\sin \theta_1}{V_{p1}} = \frac{\sin \theta_2}{V_{p2}} = \frac{\sin \phi_1}{V_{s1}} = \frac{\sin \phi_2}{V_{s2}}. \tag{4.4}$$

The other parameters in Eq. (4.3) are

$$a = \rho_2\left(1 - 2\sin^2\phi_2\right) - \rho_1\left(1 - 2\sin^2\phi_1\right); \quad b = \rho_2\left(1 - 2\sin^2\phi_2\right) + 2\rho_1\sin^2\phi_1;$$
$$c = \rho_1\left(1 - 2\sin^2\phi_1\right) + 2\rho_2\sin^2\phi_2; \quad d = 2\left(\rho_2 V_{s2}^2 - \rho_1 V_{s1}^2\right);$$
$$D = EF + GHp^2;$$
$$E = b\frac{\cos\theta_1}{V_{p1}} + c\frac{\cos\theta_2}{V_{p2}}; \quad F = b\frac{\cos\phi_1}{V_{s1}} + c\frac{\cos\phi_2}{V_{s2}};$$
$$G = a - d\frac{\cos\theta_1}{V_{p1}}\frac{\cos\phi_2}{V_{s2}}; \quad H = a - d\frac{\cos\theta_2}{V_{p2}}\frac{\cos\phi_1}{V_{s1}}.$$

(4.5)

The curves that plot $R_{pp}(\theta)$ versus the angle of incidence, θ, are called the amplitude versus offset or simply *AVO curves* (more precisely, the amplitude versus angle or *AVA curves*).

Numerous approximations to the Zoeppritz (1919) reflectivity equation have been introduced over the years (Castagna *et al.*, 1993). They are usually called *AVO approximations*. Arguably, the simplest and a very convenient one is by Hilterman (1989):

$$R_{pp}(\theta) \approx R_{pp}(0)\cos^2\theta + 2.25\Delta v\sin^2\theta = R_{pp}(0) + 2.25[\Delta v - R_{pp}(0)]\sin^2\theta;$$
$$R_{pp}(0) = (I_{p2} - I_{p1})/(I_{p2} + I_{p1}),$$

(4.6)

where $\Delta v = v_2 - v_1$ is the difference between the Poisson's ratio, v_2, of the lower half-space and that (v_1) of the upper half-space. Although approximate, Eq. (4.6) produces the reflectivity close to that produced by the exact Eq. (4.3) where the incidence angle is not large and the elastic contrast between the elastic half-spaces is small (Figure 4.2).

Because many AVO approximations employ the form where $R_{pp}(\theta)$ is a function of $\sin^2\theta$:

$$R_{pp}(\theta) = R + G\sin^2\theta,$$

(4.7)

the two parameters commonly used to describe the character of an AVO curve are the intercept, R, and gradient, G. For the Hilterman (1989) AVO approximation,

$$R = R_{pp}(0), \quad G = 2.25[\Delta v - R_{pp}(0)].$$

(4.8)

No matter whether the exact AVO equation is used or any of its approximate forms, the intercept is always

$$R = R_{pp}(0) = (I_{p2} - I_{p1})/(I_{p2} + I_{p1}) \approx 0.5\ln(I_{p2}/I_{p1}).$$

(4.9)

Table 4.1 *Elastic properties of the upper (shale) and lower (sand) half spaces used to compute the reflectivity curves in Figure 4.2. The upper half-space was shale with 65% clay and 35% quartz content and 17% porosity. The lower half-space was sand with 5% clay and 95% quartz content and porosity 25%. In the first scenario, the sand was wet. In the second and third scenarios, the sand had 40% water saturation with the rest of the pore space occupied by oil and gas, respectively. The elastic properties were computed using the Raymer–Hunt–Gardner (1980) model for V_p combined with the Dvorkin (2008a) model for V_s.*

Rock Type	V_p (km/s)	V_s (km/s)	Density (g/cm³)	I_p (km/s g/cm³)	v
Wet Shale	3.161	1.598	2.332	7.370	0.328
Wet Sand	3.645	2.041	2.235	8.145	0.271
Oil Sand	3.503	2.059	2.197	7.697	0.236
Gas Sand	3.353	2.102	2.109	7.071	0.177

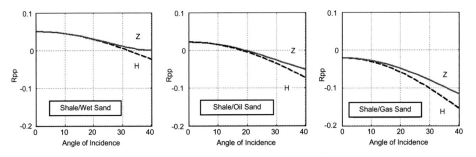

Figure 4.2 Reflectivity versus angle computed between two half-spaces using the exact Zoeppritz equation (solid curves) and Hilterman's (1989) approximation (dashed curves). Left to right: Three scenarios for wet shale over wet sand, oil sand, and gas sand. The elastic parameters used in this modeling are listed in Table 4.1. The two curves, the exact one and approximate one have the same character and are close to each other at the incidence angles below 30°, especially so for the first two scenarios.

4.3   Forward modeling using elastic constants

Because Eq. (4.6) uses only two elastic parameters, the *P*-wave impedance and Poisson's ratio, it is especially convenient for displaying the effect of the elastic constants on the AVO curve. An example of such a display is shown in Figure 4.3 where in the I_p versus v plane we first select the I_p–v pair for the upper half-space (the symbol on the right) and then for the lower half-space (the symbol on the left). From these two pairs and using Eq. (4.6), $R_{pp}(\theta)$ is computed as a function of the incident angle, and the intercept and gradient are determined from Eq. (4.8). Next, the resulting seismic gather is computed by convolving the Ricker wavelet with reflectivity determined at the interface at each incidence angle and displayed in a separate window.

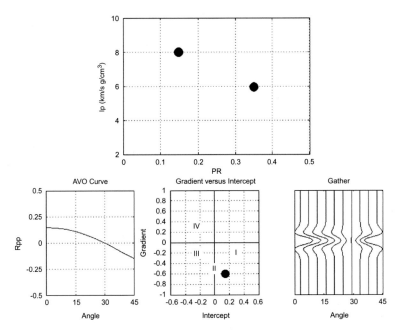

Figure 4.3 An elastic AVO modeling applet. Top: the *P*-wave impedance versus Poisson's ratio plane. Two points in this plane are selected, the first one for the upper half-space (on the right in this example) and the second one for the lower half-space (on the left). In this example, the first point is on the right-hand side of the plane while the second one is on the left-hand side of the plane and has larger impedance and smaller Poisson's ratio than the first point. Bottom, from left to right: The AVO curve; gradient versus intercept; and the seismic gather where the reflected wave traces are plotted versus the angle of incidence. The vertical axis in the gather plot is TWT (two-way travel time) or depth. The resulting AVO curve belongs to Class I.

Notice that in the example from Figure 4.3, the reflectivity is positive at zero angle of incidence. It gradually decreases as the angle increases and becomes zero at about 30°. At higher angles, it becomes increasingly negative. The gradient is negative while the intercept is positive. The transition of the AVO curve through zero amplitude at 30° is clearly discernible in the seismic gather where the phase of the trace changes at this angle of incidence. The display in Figure 4.3 is generated by an interactive code that we call an *AVO applet*.

Geophysicists often use nomenclature of four AVO types to classify the character of the AVO curve. According to Rutherford and Williams (1989) Castagna *et al.* (1998), AVO Class I is where the intercept is positive and the gradient is negative (Figure 4.3); Class II is where the intercept is near zero and the gradient is negative (Figure 4.4, top); and Class III is where both the intercept and gradient are negative (Figure 4.4, bottom). There is also AVO Class IV which is reserved for curves with positive gradient and negative intercept (Figure 4.5). These AVO classes are marked in the applet display in the gradient versus intercept plot.

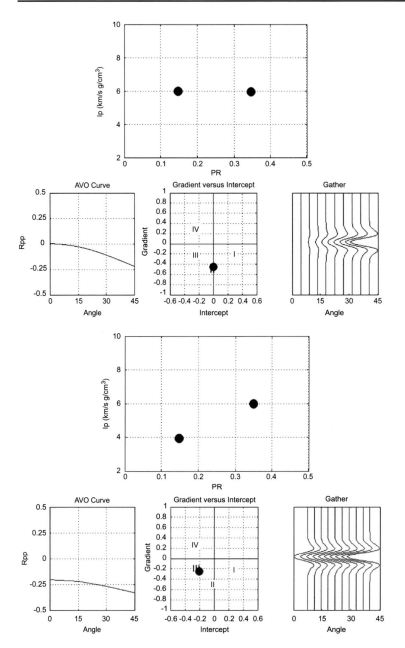

Figure 4.4 Same as Figure 4.3 but for varying *P*-wave impedance of the lower half-space: it is the same as that of the upper half-space (top applet display, AVO Class II) and is smaller than that of the upper half-space (bottom applet display, AVO Class III). In these examples, we first select the point on the right and then on the left).

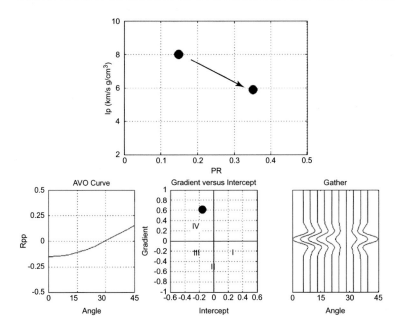

Figure 4.5 Same as Figure 4.4 (top) but with reversed order of selecting the elastic properties: here the impedance of the lower half-space is smaller than that of the upper half-space while the Poisson's ratio of the lower half-space is larger than that of the upper half-space. AVO Class IV.

AVO Class IV can occur not only for the Poisson's ratio in the lower half-space being larger than that in the upper half-space. At least theoretically, it may also be produced by a strong impedance difference between the half-spaces (Figure 4.6).

These elastic modeling results can be interpreted in terms of the physical properties of rock by using rock physics. For example, Raymer's *et al.* (1980) model combined with Dvorkin's (2008a) V_s equation predicts that wet shale with 25% porosity has $I_p = 5.42$ km/s g/cm^3 and $v = 0.40$, while gas sand with 15% porosity has $I_p = 9.81$ km/s g/cm^3 and $v = 0.14$. This contrast produces AVO Class I. Wet shale with 10% porosity has $I_p = 7.60$ km/s g/cm^3 and $v = 0.35$, while gas sand with 23% porosity has $I_p = 7.55$ km/s g/cm^3 and $v = 0.16$. This contrast produces AVO Class II. The same shale and gas sand with 30% porosity that now has $I_p = 5.88$ km/s g/cm^3 and $v = 0.19$ produces AVO Class III.

However, by using rock physics transforms, we can also forward model the AVO response directly from rock properties.

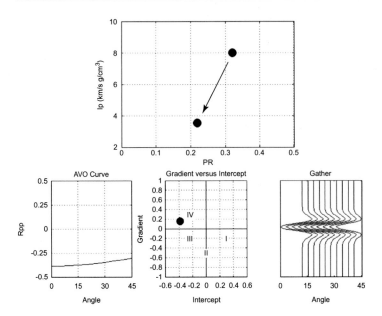

Figure 4.6 Same as Figure 4.5 but moving from a high impedance high Poisson's ratio half-space to a low impedance low Poisson's ratio half-space. AVO Class IV.

4.4   Forward modeling directly from rock properties

In the following example we assume, for simplicity, that the solid matrix of rock includes only two minerals, quartz and clay. This assumption, however, does not limit the applicability of the principles discussed here to other lithological types.

Figure 4.7 (left) shows a plane where the horizontal axis is the total porosity of rock and the vertical axis is the clay content. The two rectangles in this plot delineate the porosity and clay content ranges we select for the sand (the porosity varies between 20 and 30% while the clay content varies between 5 and 25%) and the shale (the porosity varies between 10 and 30% and the clay content varies between 75 and 95%). Let us also assume that the shale is 100% wet and its elastic properties relate to the porosity and clay content according to the Raymer *et al.* (1980) and Dvorkin (2008a) equations. Then, by applying this transform to the selected porosity and clay content ranges, we compute the respective *P*-wave impedance and Poisson's ratio and map the results onto the I_p–v plane (Figure 4.7, right). The shale rectangle on the left translates into a slanted shape on the right.

The same model is used to map the sand rectangle on the left into the sand-occupied domains in the right-hand plot. Now we can compute the respective elastic properties not only for the 100% wet sand but also for the same sand saturated with gas and oil. The slanted shapes corresponding to these three rock types are shown in the right-hand plot.

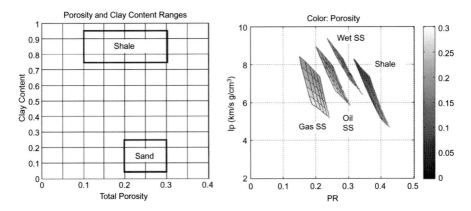

Figure 4.7 Mapping the porosity and clay content rectangles selected for shale and sand in the plot on the left onto the impedance–Poisson's ratio plane (right). The domains occupied by the shale and wet, gas, and oil sand on the right are color-coded by porosity. The higher the porosity the smaller the impedance and higher the Poisson's ratio.

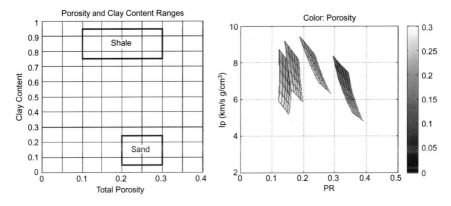

Figure 4.8 Same as Figure 4.7 but for the stiff-sand model.

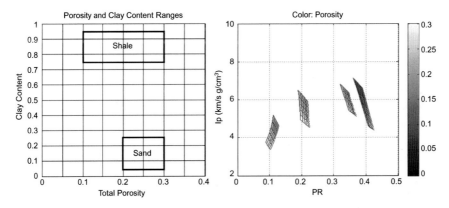

Figure 4.9 Same as Figure 4.7 but for the soft-sand model.

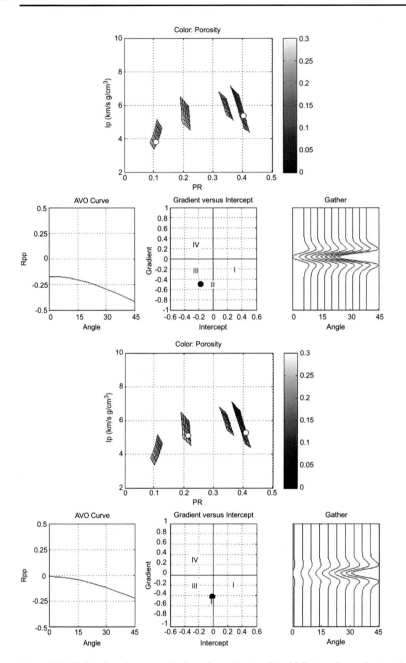

Figure 4.10 Reflections between shale and gas (top), oil (middle), and wet (bottom) sand for the soft-sand model. In the panels on the top, the shale point (symbol on the right) is selected first.

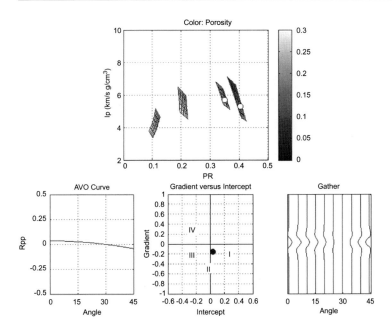

Figure 4.10 (*cont.*)

Of course, the domains occupied by shale and sand will vary depending on the rock physics transform selected. Figure 4.8 shows these domains computed for the stiff-sand model. Because the stiff-sand and Raymer *et al.* (1980) models provide very close results, the shale and sand domains in Figure 4.8 are close to those in Figure 4.7. However, the stiff-sand model results in smaller Poisson's ratio values for both sand and shale.

Figure 4.9 shows the results for the soft-sand model and using the same inputs as in the previous two examples. As expected, both shale and sand have smaller impedance compared to the stiff-sand and Raymer *et al.* (1980) models.

Now that the rock properties and fluid ranges are mapped into the elastic domain, we can create the seismic gathers from the elastic domain but directly from the rock properties. For example, by assuming that both shale and sand in the interval under examination obey the soft-sand model, we can explore how the reflections between shale and sand vary depending on the fluid (Figure 4.10). As the fluid in the sand changes from gas to oil to water, the response changes from AVO class III to AVO Class II to a very weak response with small positive intercept and small negative gradient.

Using the same applet, we can also explore how the response between shale and oil sand with 25% porosity varies depending on the degree of consolidation of shale by gradually varying the shale's porosity from high 30% to intermediate 20% and to small 10% (Figure 4.11). The respective response varies from a weak AVO Class I to AVO Class II to AVO Class III.

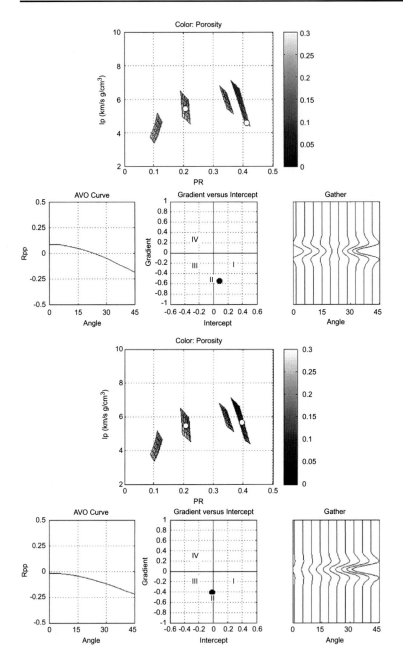

Figure 4.11 Reflections between shale and oil sand as the shale becomes more consolidated (top to bottom) for the soft-sand model.

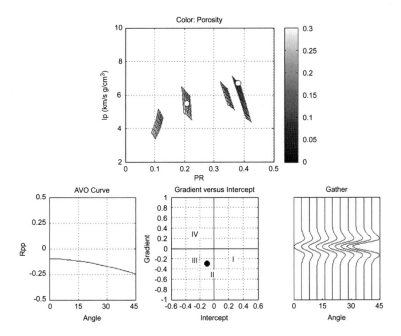

Figure 4.11 (*cont.*)

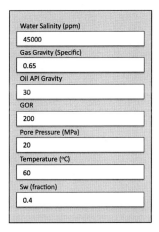

Figure 4.12 Example of input table of fluid properties and conditions for AVO modeling shown in previous figures of this section.

The rock physics model appropriate for a specific interval under examination is generally a priori unknown. To establish such a model, we need to conduct rock physics diagnostics (Chapter 3). Then the appropriate fluid properties and conditions have to be used as the input for applet modeling (Figure 4.12). Finally, forward modeling for all desirable scenarios can be conducted as shown in Figures 4.7 to 4.11.

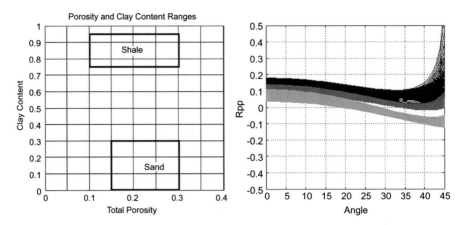

Figure 4.13 AVO curves (right) at the interface between shale and wet (black), oil (dark gray), and gas (light gray) sand using the inputs shown in Figure 4.12 and sand and shale properties delineated on the left.

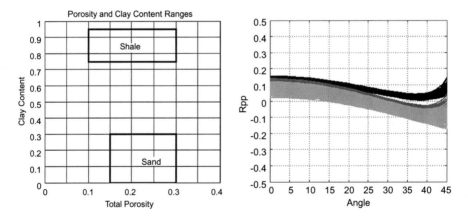

Figure 4.14 Same as Figure 4.13 but for the stiff-sand model.

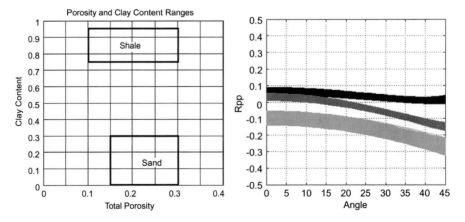

Figure 4.15 Same as Figure 4.13 but for the soft-sand model.

Another type of rock physics applet uses the same input dialogue as shown in Figure 4.12 and the sand and shale property range selection as shown in Figure 4.7 (left). Following this parameter setup, we compute the AVO curves between shale and wet, oil, and gas sand, this time using the full Zoeppritz equations. Figures 4.13, 4.14, and 4.15 show the result where both sand and shale obey the Raymer *et al.* (1980) and Dvorkin (2008a) models.

5 Pseudo-wells: principles and examples

5.1 Sandwich (three-layer) model

The simplest configuration for a pseudo-well is a three-layer sandwich where a finite-thickness reservoir is bounded by two non-reservoir layers, for example, sand embedded in shale. The properties of shale and sand have to be selected to reflect the relevant geological conditions at the site of investigation. The same holds for the conditions that include the pore and differential pressure and fluid properties.

For the examples shown in this section, we select the fluid properties and conditions listed in Table 5.1. Also, in order to translate the rock properties and conditions into the elastic properties, we use the soft-sand model where the differential pressure is 25 MPa, critical porosity is 0.40, coordination number is 6, and the shear modulus correction factor, f, (Eq. (2.26)) is 1. The density and elastic properties of the fluid phases are computed using the Batzle–Wang (1992) equations. Finally, the effective bulk modulus of mixtures of brine and oil and brine and gas are computed using the harmonic average of the bulk moduli of the respective fluid phases (Eq. (2.11)). An example of the shale and sand elastic properties according to the soft-sand model in the impedance–porosity and impedance–Poisson's ratio planes is shown in Figure 5.1.

In our first example, we assume that the sand reservoir contains 40% brine and 60% oil; the clay content in the shale is 80% and its porosity is 0.20; the clay content in the sand is 5%; and the underlying shale is the same as the shale above the reservoir. We gradually reduce the porosity of the sand from 35% to 15% and produce the respective synthetic gathers (Figure 5.2). The seismic response is computed using a ray tracer with full Zoeppritz equations and a Ricker wavelet with a 60 Hz center frequency. The same seismic parameters were used in all examples in this section.

The resulting seismic gathers indicate that the strong AVO Class III response for high-porosity sand becomes weaker as the sand's porosity decreases from 35% to 25% (the upper three graphs in Figure 5.2). When the sand's porosity is 20%, the response becomes AVO Class II (the fourth from top graph in Figure 5.2). Finally, it becomes AVO Class I when the sand's porosity is 15% (the bottom graph in Figure 5.2).

Table 5.1 *Fluid properties and conditions used in computations for Figures 5.1 to 5.5.*

Brine Salinity	Oil API	Gas Gravity	GOR	Pore Pressure (MPa)	Temperature (°C)
50,000	30	0.65	300	25	70

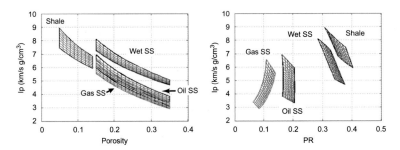

Figure 5.1 Soft-sand model. Impedance versus porosity (left) and impedance versus Poisson's ratio (right) computed for shale whose porosity varies between 0.05 and 0.14 and the clay content between 60 and 100% and sand whose porosity varies between 0.15 and 0.35 and the clay content between zero and 30%. In the oil sand the water saturation is 40% while in the gas sand it is 20%. The impedance values of the oil and gas sand overlap. The domains occupied by the gas sand are outlined by gray curves. In the plot on the left, the impedance decreases with the increasing clay content. In the plot on the right, the sub-vertical curves are for constant clay content while sub-horizontal curves are for constant porosity and varying clay content.

The model used to produce the synthetic gathers shown in Figure 5.2 can be utilized to explore, for example, the influence of changing oil properties on the seismic response. In the example illustrated in Figure 5.3, we fix the porosity and clay content of the shale as 0.20 and 80%, respectively, while the porosity and clay content in the sand are 0.25 and 5%, respectively. The water saturation in the sand is 40%. The variable in this example is GOR, which gradually decreases from high 600 to zero (dead oil). The input rock properties and saturation are shown in the first three tracks in Figure 5.3. The results show the transition from a strong AVO Class III response for high GOR to a weaker AVO Class III response to a weak AVO Class I response for the dead oil.

In our next example (Figure 5.4) we explore the effect of the rising water table in gas sand whose porosity is 0.30 and water saturation is 20%. All other parameters are the same as in the previous examples. As the water table gradually rises, the originally strong AVO Class III response becomes weaker and eventually becomes an AVO Class II response for wet sand.

The above three examples illustrate the non-uniqueness of the seismic response: the change in AVO class may be triggered by variations in porosity, oil properties, or the position of the water table. To mitigate this uncertainty, the input has to be selected to represent the site-specific properties and the variable whose effect is investigated has to be selected based on the most likely geologic scenarios at the site.

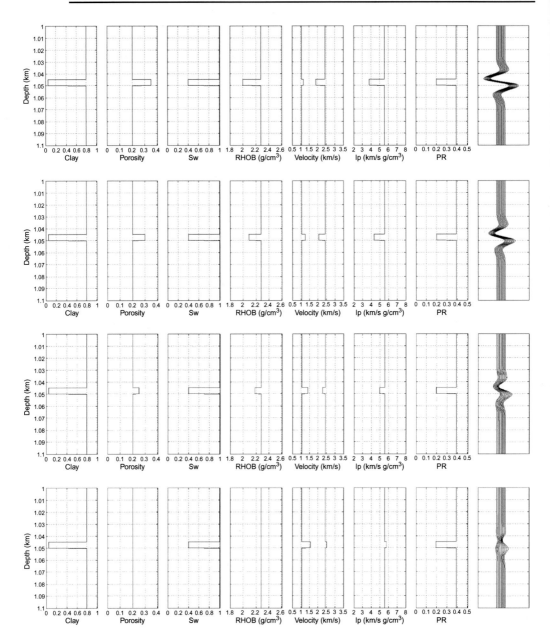

Figure 5.2 Synthetic gathers for a sandwich model. Effect of varying porosity of the oil sand. The input properties of shale and sand are displayed in the first three tracks in each graph. The elastic properties computed using the soft-sand model are displayed in the 4th, 5th, 6th, and 7th tracks. The resulting synthetic seismic gathers are displayed in the 8th track. The trace on the left is for the normal incidence. The rest are for gradually increasing offset (left to right).

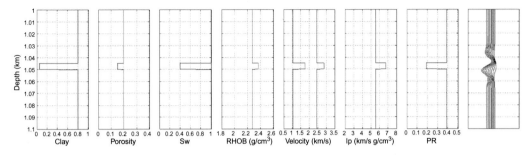

Figure 5.2 (*cont.*)

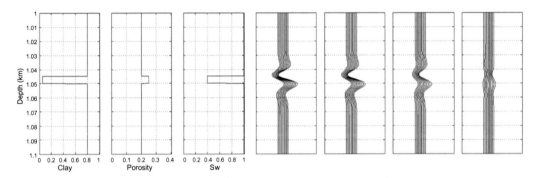

Figure 5.3 Synthetic gathers for a sandwich model. Effect of varying GOR. The input properties of shale and sand are displayed in the first three tracks in each graph. The elastic properties computed using the soft-sand model are not displayed. The resulting synthetic seismic gathers are displayed in the 4th, 5th, 6th, and 7th tracks for GOR 600, 400, 200, and zero, respectively. The trace display is the same as in Figure 5.2.

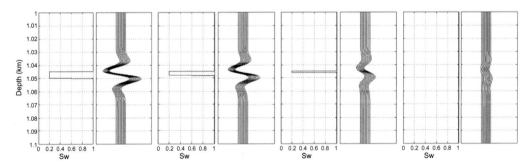

Figure 5.4 Synthetic gathers for a sandwich model. Effect of rising water table. The geometry and rock properties are the same as used in Figure 5.2. The sand's porosity is fixed at 0.30. However, here the lower part of the reservoir is wet. The location of the water table is illustrated by the water saturation curves in the 1st, 3rd, 5th, and 7th tracks. The respective synthetic gathers are shown in the 2nd, 4th, 6th, and 8th tracks. The first two tracks correspond to the case where the entire reservoir is occupied by gas. In the following tracks, the water table rises from ½ to ¾ of the sand's thickness. In the final two tracks, the entire sand is wet.

Our next example is for the so-called "wedge" model (Figure 5.5) where the thickness of the reservoir is gradually reduced and corresponding synthetic traces are computed. In this specific example, we use the same inputs as in the previous examples and fix the porosity of the gas sand at 0.30 and water saturation at 20%.

The synthetic seismic response for the sand thickness 10, 5, 2.5, and 0.5 m is shown in Figure 5.6. The strong AVO Class III response at the 10 m thick sand becomes even stronger at the 5 m thick sand due to the so-called "tuning," where the superimposed reflections from the top and bottom of the sand reservoir enhance the summary response. Thereafter, the smaller the thickness of the sand the weaker the seismic response.

Using the same principles, we can explore a double-sandwich model where two separate layers of sand are surrounded by shale. In the example shown in Figure 5.7, where the thickness of each gas sand layer is 2.5 m, porosity is 0.30, and water saturation is 20%. The distance between the top of the upper layer and the top of the lower layer gradually decreases from 20 to 10 to 5 to 0.5 m. As the distance between the sand layers becomes smaller, the separate reflections merge to form a stronger AVO Class III response.

5.2   Geologically consistent inputs

The rock properties and conditions used in the previous section to produce a pseudo-well are often interrelated and do not vary independently of each other. One such example comes from the well data discussed in Chapter 3 (Figure 3.1). The well data and their cross-plots are shown in Figure 5.8.

The porosity versus GR cross-plot in Figure 5.8 indicates that as GR increases, the porosity of the sediment decreases. This means that assuming that GR is a proxy for shale (or clay) content, the latter and the porosity in this well are connected to each other: the porosity is about 0.23 in the relatively clean sand with a GR of about 30 and decreases to about 0.05 as the GR increases from about 30 to 75. Remarkably, water saturation is also a function of GR (or shale or clay content): it is about 20% at the lowest GR point and steadily increases reaching 100% at GR $\approx$ 80.

Another example from a shallow section of an offshore well is shown in Figure 5.9. The interval under examination is fully water-saturated and, hence, here we cannot observe how saturation varies with the clay content. Still, we see a distinct V-shaped dependence of porosity and GR, qualitatively similar to that exhibited in Figure 5.8.

Figure 5.10 displays the depth plots of GR, porosity, and water saturation in the upper pay zone of another offshore gas well drilled through a sand/shale sequence. The V-shaped pattern exhibited in Figures 5.8 and 5.9 is barely present here. Instead, we see a gradual reduction in porosity with increasing GR. This is because the shale in this interval is more compacted than in the examples shown in the previous two figures so

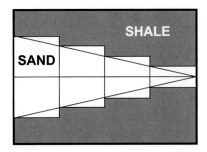

Figure 5.5 A sand wedge within shale. The seismic response is computed for gradually decreasing thickness of the sand body.

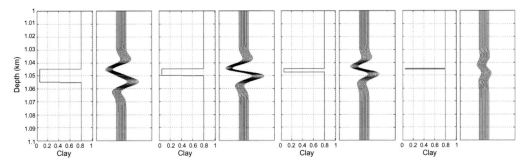

Figure 5.6 Synthetic gathers for a wedge model. The water saturation in the gas sand is fixed at 20%. The porosity of the sand is fixed at 0.30. The thickness of the sand (left to right) is 10, 5, 2.5, and 0.5 m.

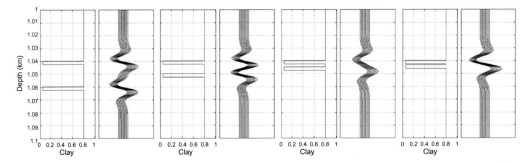

Figure 5.7 Synthetic gathers for a double-sandwich model. The water saturation in the gas sand is fixed at 20%. The porosity of the sand is fixed at 0.30. The distance between the top of the upper layer and top of the lower layer gradually decreases from 20 to 10 to 5 to 0.5 m.

that the porosity of the high-GR "pure-shale" member of the sequence is smaller than that in the shale with lower GR values, yet the sand still retains fairly high porosity. In the same figure we can also observe how water saturation steadily increases with the increasing shale content (increasing GR).

Figure 5.11 shows the same curves and cross-plots as Figure 5.10 but for the lower pay zone of the same well. The porosity–GR behavior in this interval is similar to

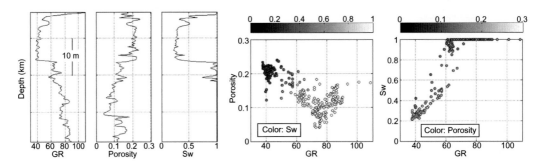

Figure 5.8 An oil well drilled through a sand/shale sequence (also discussed in Chapter 3). From left to right: First three plots are GR, porosity, and water saturation versus depth. The 4th plot is porosity versus GR, color-coded by water saturation. The 5th plot is water saturation versus GR, color-coded by porosity.

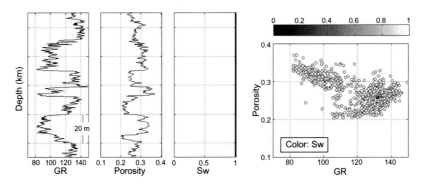

Figure 5.9 A wet interval in an offshore well drilled through a sand/shale sequence. The display is the same as in Figure 5.8. Because water saturation is 100%, the saturation versus GR plot is not shown.

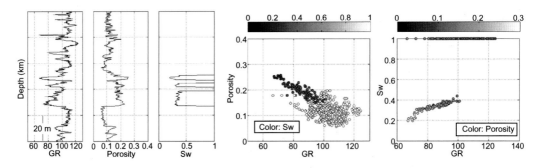

Figure 5.10 An offshore gas well drilled through a sand/shale sequence. Upper reservoir. The display is the same as in Figure 5.8.

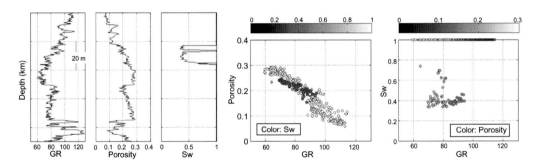

Figure 5.11 Lower reservoir in the well shown in Figure 5.10. The display is the same as in Figure 5.8.

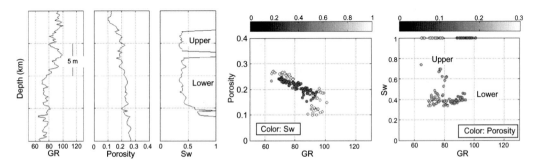

Figure 5.12 Zoom on the reservoir zone in the interval displayed in Figure 5.11. The two saturation versus GR trends shown in the 5th plot correspond to the upper and lower gas-saturated zones as marked in the 3rd graph.

that exhibited in Figure 5.10. However, in the saturation–GR cross-plot two trends are present. The steep trend comes from the lower gas sand while the relatively flat trend, similar to that shown in Figure 5.11 comes from the upper gas sand (Figure 5.12).

These examples teach us that although we often see natural interdependence of porosity, lithology, and water saturation, the trends may vary not only between two different geographical locations but even at the same location depending on the depositional characteristics of an interval selected. Yet, the existence of these trends has a robust physical foundation.

Consider a mixture of grains with two very different sizes: the larger grains represent sand while the smaller grains represent shale (or clay). At zero shale (clay) content, the sediment is made of relatively large sand grains with the total porosity, ϕ, equal that of the pure-sand end-member ϕ_{SS} (Figure 5.13). At 100% shale (clay) content, the total porosity is that of the pure-shale (clay) end-member, ϕ_{SH}. Let us assume that as the shale (clay) content increases, the original framework of large sand grains remains undisturbed and the small shale particles gradually fill the pore space of this framework. Because the porosity of the pure shale is larger than zero, even when the

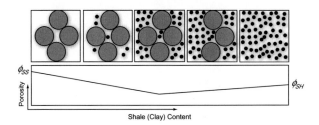

Figure 5.13 Ideal binary mixture model (top) with a schematic porosity versus clay content graph (bottom).

entire pore space of the pure-sand framework is filled with the shale particles, the total porosity is larger than zero and equals $\phi_{SS}\phi_{SH}$. As the shale content increases past this point, the pure-sand framework has to change to accommodate additional shale particles. Now we are dealing with sand grains suspended in the pure-shale frame-work. Hence, the fewer sand grains we have in this suspension the larger its porosity, which now gradually increases from its turning-point value, $\phi_{SS}\phi_{SH}$, to ϕ_{SH} at 100% shale content.

This idealized representation of sand/shale mixtures is called the ideal binary mixture model. Its rock physics implementation is called the Thomas-Stieber model. This *dispersed-shale* model is discussed in detail in Mavko *et al.* (2009). As illustrated by the well data examples given in this section, this idealized model sometimes works and, hence, is a useful approximation for the observed behavior of highly complex sediment. Notice that in this discussion we interchangeably use terms "clay" and "shale." The former usually refers to the mineralogy of the sediment while the latter refers to the grain size. Hence, in the context of the ideal binary mixture model, the term "shale" may be more appropriate. However, the clay particles are very small, thus allowing us to apply this model to sand/clay mixtures.

Following this model, we can derive simple equations for the dependence of the total porosity, ϕ, on the clay content, C (Mavko *et al.*, 2009). Specifically, for $0 \leq C \leq \phi_{SS}$,

$$\phi = \phi_{SS} - (1 - \phi_{SH})C, \tag{5.1}$$

while for $\phi_{SS} \leq C \leq 1$,

$$\phi = \phi_{SH}C, \tag{5.2}$$

where C is the volume fraction of microporous dispersed shale (clay) in the rock. Notice also that both equations give $\phi = \phi_{SS}\phi_{SH}$ at $C = \phi_{SS}$. Figure 5.14 (left) displays the V-shaped porosity versus clay content trends for varying porosity of the pure-shale end-member.

A different porosity versus clay content behavior is usually observed in a so-called *laminated* system (this is a rock physics term which a geologist may call *thinly inter-*

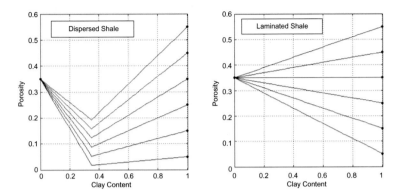

Figure 5.14 Porosity versus clay content for dispersed (left) and laminated (right) sand/shale mixtures. The porosity of the pure-sand member is fixed at 0.35 while the pure-shale porosity varies between 0.05 and 0.55.

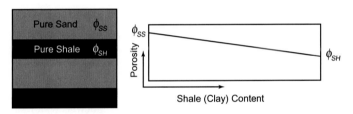

Figure 5.15 Laminated sand/shale sequence with a linear monotonic behavior of the total porosity versus clay content.

bedded) where, in the simplest case, an interval comprises alternating pure-sand and pure-shale layers where the former has porosity ϕ_{SS} while the latter has porosity ϕ_{SH} (Figure 5.15). In this case

$$\phi = \phi_{SS}(1-C) + \phi_{SH}C. \tag{5.3}$$

It is assumed that the layers are thin so that the well-log tool samples the effective rock properties of the entire laminated sequence. The respective ϕ versus C trends are shown in Figure 5.14 (right).

To explain the saturation versus clay content dependence, let us remember that hydrocarbon cannot enter an originally water-saturated system if its irreducible water saturation, S_{wi}, is 100%. S_{wi} depends on the magnitude of capillary forces inside porous rock: if these forces are large, the water simply cannot be displaced and, hence, $S_{wi} = 1$.

The smaller the grains (at fixed porosity) the larger the capillary forces. This is why the shale (barred gas and oil shale in which hydrocarbons are generated) is usually 100% wet. The same is true for sand with a large clay content. By, once again, resorting to idealized models and equations, we can describe the observed S_w versus C behavior.

Let us first assume that water saturation in a reservoir is always at its irreducible saturation value (of course this assumption is often violated in real sediment with hydrocarbons). Let us recall next one of the empirical permeability equations that relate the absolute permeability, K, to the total porosity, ϕ, and irreducible water saturation, S_{wi}:

$$k = 8581\phi^{4.4} / S_{wi}^2, \qquad (5.4)$$

which is called Timur's equation (Mavko *et al.*, 2009), where the permeability is in milliDarcy (mD) and both porosity and water saturation are unitless volume fractions. Let us next recall the Kozeny–Carman permeability equation (Mavko *et al.*, 2009) that relates the absolute permeability to the grain size, d:

$$k = d^2 \frac{10^9}{72} \frac{(\phi - \phi_p)^3}{[1 - (\phi - \phi_p)]^2 \tau^2}, \qquad (5.5)$$

where ϕ_p is the percolation (or threshold) porosity at which the pore space becomes disconnected and, hence, the permeability becomes zero; τ is the unitless tortuosity; and d is in mm. A reasonable range for ϕ_p is between zero and 0.03. The tortuosity for medium-to-high porosity sandstone is between 2.0 and 3.0.

By combining Eqs (5.4) and (5.5) we can relate the irreducible water saturation to porosity, grain size, and tortuosity as

$$S_{wi} = \frac{0.025}{d} \frac{\phi^{2.2}[1 - (\phi - \phi_p)]\tau}{(\phi - \phi_p)^{1.5}}, \qquad (5.6)$$

where the units are the same as in Eqs (5.4) and (5.5). For $\phi_p = 0$, Eq. (5.6) reads

$$S_{wi} = 0.025 \frac{\phi^{0.7}(1 - \phi)\tau}{d}. \qquad (5.7)$$

For a porous system with mixed particle sizes, the effective particle size that can be used in the Kozeny–Carman equation is the harmonic average of the individual particle sizes (Mavko *et al.*, 2009). Specifically, if the particle size in the sand is d_{SS} and that in the shale (clay) is d_{SH}, the effective particle size d is

$$d = \left(\frac{1-C}{d_{SS}} + \frac{C}{d_{SH}} \right)^{-1}. \qquad (5.8)$$

The range of the grain sizes in sand is between 0.050 and 2.000 mm, it is between 0.002 and 0.050 mm in silt, and less than 0.002 mm in clay. Figure 5.16 (left) shows

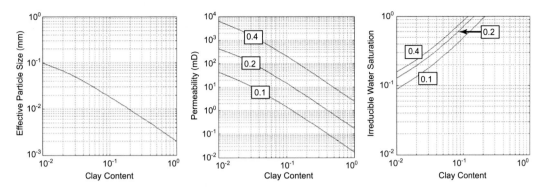

Figure 5.16 Left: The effective grain size versus the clay content according to Eq. (5.8) and for the sand grain size 0.200 mm and shale grain size 0.002 mm. Middle: Permeability versus the clay content according to Eq. (5.5) with the percolation porosity zero, tortuosity 2, and the same sand and shale grain sizes. Permeability is computed for porosity 0.4, 0.2, and 0.1 as marked in the plot. Right: The irreducible water saturation versus the clay content for the same inputs and three porosity values.

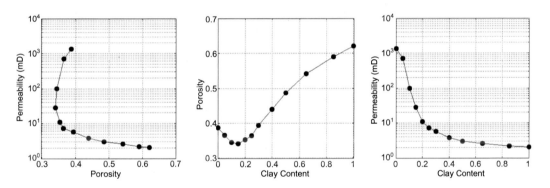

Figure 5.17 Ottawa sand and kaolinite mixtures (Yin, 1992). Left: Permeability versus porosity. Middle: Porosity versus clay content. Right: Permeability versus clay content.

how the effective grain size varies with the shale (clay) content for $d_{SS} = 0.010$ mm and $d_{SH} = 0.001$ mm.

The drastic permeability reduction with the increasing clay content is illustrated by the experimental data from Yin (1992) obtained on mixtures of Ottawa sand and kaolinite clay (Figure 5.17). These data were obtained at room conditions at which the porosity of pure kaolinite is very high, in excess of 0.6. As the clay content steadily increases from zero to 100%, we first observe a porosity reduction from that of the pure sand (about 0.40) to its minimum value (about 0.35) followed by monotonic porosity increase to that of the pure clay. This is why in the permeability–porosity plot in this figure we observe a somewhat counterintuitive reduction of permeability with increasing porosity.

Let us now return to our analytical derivations and combine Eqs (5.6) and (5.8) to relate the irreducible water saturation to the clay content as

$$S_{wi} = 0.025\left(\frac{1-C}{d_{SS}} + \frac{C}{d_{SH}}\right)\frac{\phi^{2.2}[1-(\phi-\phi_p)]\tau}{(\phi-\phi_p)^{1.5}} \tag{5.9}$$

or

$$S_{wi} = 0.025\phi^{0.7}(1-\phi)\tau\left(\frac{1-C}{d_{SS}} + \frac{C}{d_{SH}}\right) \tag{5.10}$$

for $\phi_p = 0$.

Let us once again fix the sand and shale grain sizes as $d_{SS} = 0.200$ mm and $d_{SH} = 0.002$ mm and also assume $\tau = 2$ and $\phi_p = 0$. The resulting permeability and irreducible water saturation according to Eqs (5.5) and (5.10), respectively, are plotted versus the clay content for several porosity values in Figure 5.16.

Notice that the analytical derivations that gave rise to the curves plotted in Figure 5.16 are based on combining an empirical Eq. (5.4) with idealized Eqs (5.5) and (5.8). This means that caution has to be exercised when applying these closed-form equations to a real case. Still, in the absence of site-specific relations for irreducible water saturation, Eq. (5.9) can be used as an approximate estimate for S_{wi}. Remember also that because some sands simply do not have hydrocarbons, even if the irreducible water saturation is less than 1, the actual water saturation can be 100% or anywhere between S_{wi} and 100%.

Let us finally combine Eq. (5.1) that relates the total porosity to the clay content for the dispersed sand/shale system with the irreducible water saturation equation to directly relate S_{wi} to C and obtain

$$S_{wi} = 0.025\left(\frac{1-C}{d_{SS}} + \frac{C}{d_{SH}}\right)\frac{[\phi_{SS} - (1-\phi_{SH})C]^{2.2}(1-\{[\phi_{SS} - (1-\phi_{SH})C] - \phi_p\})\tau}{\{[\phi_{SS} - (1-\phi_{SH})C] - \phi_p\}^{1.5}} \tag{5.11}$$

for $0 \le C \le \phi_{SS}$, or

$$S_{wi} = 0.025[\phi_{SS} - (1-\phi_{SH})C]^{0.7}\{1 - [\phi_{SS} - (1-\phi_{SH})C]\}\tau\left(\frac{1-C}{d_{SS}} + \frac{C}{d_{SH}}\right) \tag{5.12}$$

for $0 \le C \le \phi_{SS}$ and $\phi_p = 0$.

Figure 5.18 shows the results of using Eq. (5.12) with $\phi_{SS} = 0.35$, $d_{SS} = 0.200$ mm, $d_{SH} = 0.002$ mm, $\tau = 2$, $\phi_p = 0$, and $\phi_{SH} = 0.55, 0.25$, and 0.05. We notice that S_{wi} is practically independent of the porosity of the pure-shale member and it rapidly approaches 1 as the clay content increases from zero to about 0.15.

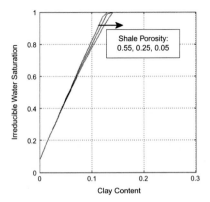

Figure 5.18 Irreducible water saturation versus the clay content according to Eq. (5.12) for varying porosity of the pure shale member.

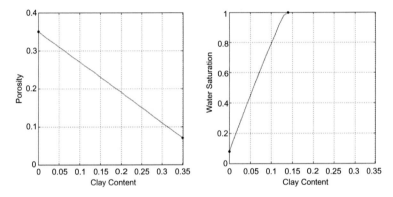

Figure 5.19 The total porosity (left) and water saturation (right) in the sand as its clay content increases from zero to 0.35.

Now we can apply the equations that link the sand's clay content to its total porosity and water saturation. Specifically, we start with an unconsolidated sand layer with porosity 0.35 and zero clay content embedded in shale with porosity 0.20. The soft-sand model is used for computing the elastic properties of sand and shale. The clay content in the sand gradually increases from zero to 0.35. We use Eq. (5.1) to link the sand's porosity to its clay content. We also assume that the water saturation is the irreducible water saturation, S_{wi}, as given by Eq. (5.12) where the inputs are chosen as $d_{SS} = 0.200$ mm, $d_{SH} = 0.002$ mm, $\tau = 2$, and $\phi_p = 0$. The porosity of the pure-sand end-member is $\phi_{SS} = 0.35$ and that of the pure-shale member is $\phi_{SH} = 0.20$.

Figure 5.19 shows how the total porosity and water saturation in the sand vary with the clay content. Figure 5.20 shows the computed porosity, water saturation, the elastic properties, and synthetic gather for selected clay content scenarios. Figure 5.21 shows the synthetic gathers computed for the entire clay content range in the sand layer.

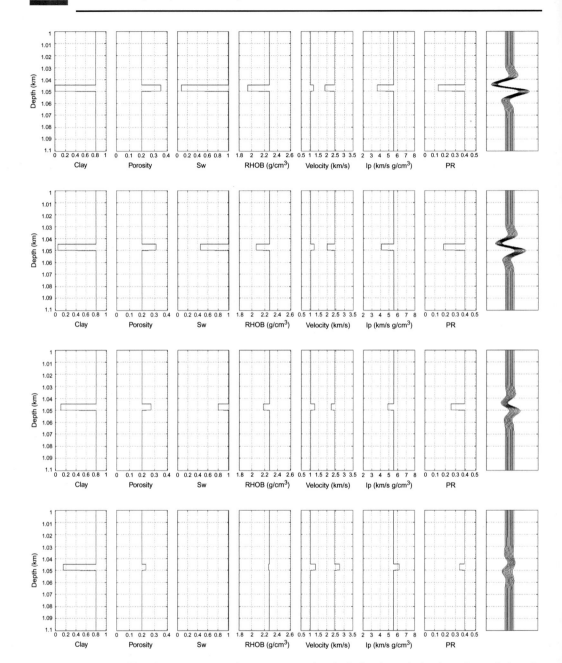

Figure 5.20 The clay content, porosity, water saturation, bulk density, velocity, impedance, Poisson's ratio, and the resulting synthetic seismic gather (left to right) as the clay content in the sand changes from zero to 5, 10, and 15% (top to bottom).

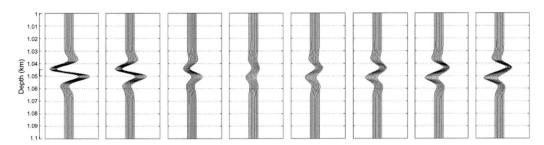

Figure 5.21 Seismic gathers at a sand layer as the clay content increases from zero to 35% with 5% steps (left to right).

Figure 5.20 shows that as we increase the clay content in the sand, its porosity gradually decreases and the water saturation increases. In this example, the sand becomes fully wet at about 13% clay content. The AVO type is a strong Class III in clean sand and becomes weaker as we add more clay. For the clay content of zero, 5, and 10%, the impedance in the sand layer exceeds that in the overburden shale. When the sand becomes fully wet, its impedance exceeds that in the shale and the AVO response becomes a weak Class I. Adding more clay to the sand acts to reduce its porosity and increase the positive impedance contrast between the sand and overburden shale (Figure 5.21).

If we did not vary the porosity and water saturation in the sand as the clay content increases, its impedance would become smaller and smaller (the higher the clay content at fixed porosity the softer the rock). The AVO response would remain Class III and become stronger and stronger with the increasing clay content.

This example shows the importance of modeling the rock properties in a depositionally consistent way. In this example we used highly idealized analytical expressions based in part on an empirical equation (Eq. (5.4)) which is specific to the dataset used to obtain this equation. Still, these equations linking the total porosity and water saturation to the clay content can be used for rough estimates of the seismic response.

5.3   Depth and compaction

When constructing a pseudo-well, it is important to remember that the porosity of the sediment, especially that of shale, usually decreases with increasing depth. This is due to the load imposed by the overburden, which monotonically increases with depth. There is no universal formula for the porosity–depth relations: they usually vary from basin to basin and even within the same basin. Porosity reduction in shale is commonly much more dramatic than in sand, as exemplified by Yin's (1992) experiments on compacting dry mixtures of Ottawa sand and kaolinite clay (Figure 5.22).

Mechanical compaction is not the only porosity reduction mechanism. In addition, chemical compaction (diagenesis) can take place in sand in certain pressure and

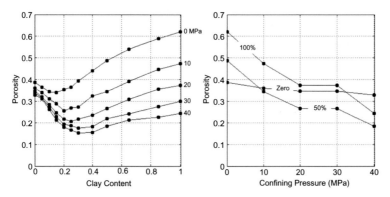

Figure 5.22 Ottawa sand and kaolinite mixtures (Yin, 1992). Left: Porosity versus clay content at confining pressure zero, 10, 20, 30, and 40 MPa. Right: Porosity versus confining pressure for the samples with zero, 50, and 100% clay content. Compare the end-member behavior in the left-hand plot: while the pure-sand porosity decreases from 0.387 to 0.329 as the confining pressure is raised from zero to 40 MPa, the pure-shale porosity decreases from 0.621 to 0.244.

temperature windows, resulting in the dissolution and precipitation of quartz and the subsequent cementation of the grains (Avseth *et al.*, 2005). As a result, the porosity reduces beyond the limit afforded by mechanical grain rearrangement, and this is accompanied by a strong increase in the elastic-wave velocity.

In shale, the degree of compaction depends on the difference between the overburden stress and the pore pressure. In many basins and at certain depth, the pore pressure can be abnormally high, higher than the hydrostatic pressure exerted by a water column. Such overpressured shales usually have a higher porosity than normally pressured shales at smaller depths, which is accompanied by a reduction of the elastic-wave velocity. In the overpressure example illustrated by Figure 5.23, the shale porosity decreases with depth in the upper part of the interval and then starts to increase with depth. The porosity in the gas sand (the low-GR feature at the bottom) is much higher than that in the overlying shale.

Published equations for laboratory- and field-based compaction in shale and sand reflect attempts to generalize the observed diversity of porosity reduction behavior with depth. Based on a common observation that the rate of porosity reduction is rapid at shallow depth and slows at greater depth of burial, Athy (1930) used an exponent to describe porosity collapse with depth:

$$\phi = \phi_0 e^{-cZ}, \tag{5.13}$$

where ϕ_0 is the porosity of the uncompacted material, Z is depth, and c is a positive constant measured in Z^{-1} (km^{-1} in the units used here). Another functional form cited by Schon (2004) is logarithmic:

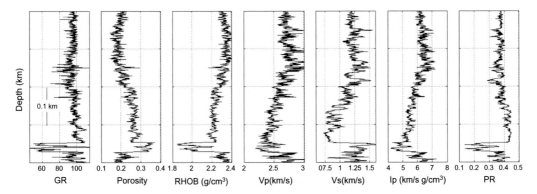

Figure 5.23 Depth plots in a well with overpressure. The porosity in the shale (2nd track) decreases with depth in the first 100 m of the interval and then starts to gradually increase due to overpressure. The overpressured shale is located directly above the gas sand interval manifested by low GR and high porosity. The bulk density (3rd track) increases with depth in the upper part of the interval and then starts to gradually decrease due to overpressure. The velocity and impedance behave similarly. The Poisson's ratio in the overpressured shale is slightly higher than in the normally pressured shale in the upper portion of the interval. It is low in the gas sand.

$$\phi = \phi_0 - A \ln(Z), \tag{5.14}$$

where A is a positive constant.

Schon (2004) also cites the following empirical equations for the compaction trends:

$$\phi = \phi_0 e^{-0.45Z} \tag{5.15}$$

for the Russian Platform sediments;

$$\phi = 0.496 e^{-0.556Z} \tag{5.16}$$

for sandstones in Yugoslavia; as well as equations for the Northern North Sea sands and shales:

$$\phi = 0.49 e^{-0.27Z} \tag{5.17}$$

for sandstone and

$$\phi = 0.803 e^{-0.51Z}, \tag{5.18}$$

for shale, where the depth Z is in km.

Site-specific compaction trends have to be calibrated to well data. Consider for example an offshore well with the GR and total porosity profiles shown in Figure 5.24.

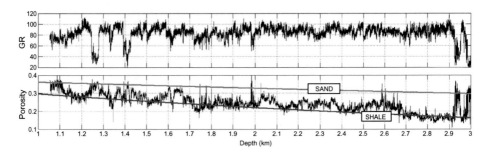

Figure 5.24 An offshore well with compacting sand and shale. Depth is the horizontal axis. GR (top) and porosity (second from top) versus depth. The lower curve in the porosity plot is according to Eq. (5.19) while the upper curve is according to Eq. (5.20).

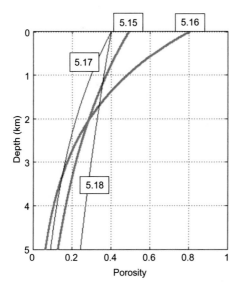

Figure 5.25 Porosity compaction trends given by Eqs (5.15), (5.16), (5.17), and (5.18) as marked in the plot.

The interval under examination spans a depth range between 1 and 3 km where the total porosity of the shale is significantly smaller than that of the sand. The shale porosity reduction trend can be approximated as

$$\phi_{SHALE} = 0.40e^{-0.30 \times \text{Depth}}, \tag{5.19}$$

while that in the sand is approximately

$$\phi_{SAND} = 0.40e^{-0.10 \times \text{Depth}}. \tag{5.20}$$

These trends are different from those given by Eqs (5.17) and (5.18) (Figure 5.25) likely because this well was drilled at a geographically different location where the depositional regime is not the same as in the Northern North Sea.

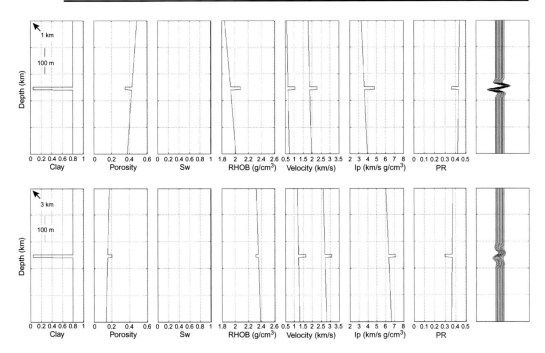

Figure 5.26 Wet sand. Rock conditions and computed elastic properties and seismic gathers at 1.25 (top) and 3.25 (bottom) km depth.

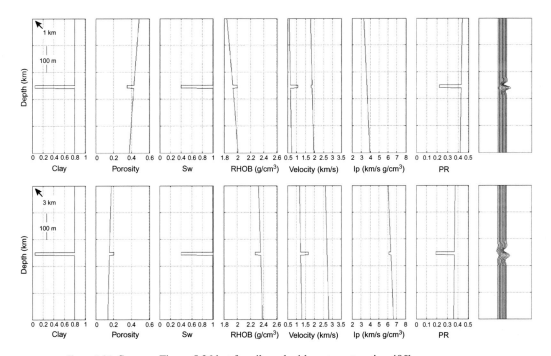

Figure 5.27 Same as Figure 5.26 but for oil sand with water saturation 40%.

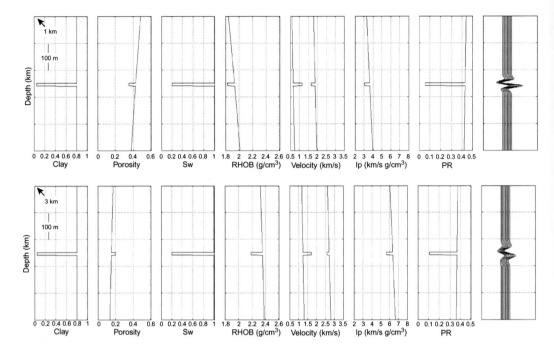

Figure 5.28 Same as Figure 5.26 but for gas sand with water saturation 20%.

Let us next adopt Eqs (5.17) and (5.18) to create a pseudo-well where two separate sand layers are located at different depths (Figure 5.26). Assume also that the clay content in the shale is 80% while that in the sand is 5% and the elastic properties of both sand and shale obey the soft-sand model. The properties of the reservoir fluids are given by Table 5.1, except that in our computations of the density and bulk modulus we will use the pore pressure, P_{PORE}, increasing with depth due to the increasing weight of the brine column:

$$P_{PORE} = 10\rho_w Z, \tag{5.21}$$

where pressure is in MPa and ρ_w is the density of water assumed to be 1.0 g/cm^3. The temperature, T, is also increasing with depth starting at 5° C at the mudline and with the geothermal gradient $G_T = 20°$ C per km:

$$T = 5 + G_T Z. \tag{5.22}$$

The differential pressure to be used in the soft-sand model is the difference between the overburden pressure, P_{OVER}, and the pore pressure. P_{OVER} has to be computed as an integral of the bulk density with respect to depth:

$$P_{OVER} = 10 \int_0^Z \rho dz. \tag{5.23}$$

Let us assume for simplicity that the average bulk density is 2.3 g/cm³. Then

$$P_{OVER} = 10\rho_b Z = 23Z, \tag{5.24}$$

which means that the differential pressure P_{DIFF} in the following example is

$$P_{DIFF} = P_{OVER} - P_{PORE} = 13Z. \tag{5.25}$$

We examine three pore-fluid scenarios: wet sand (Figure 5.26), oil sand with 40% water saturation (Figure 5.27), and gas sand with 20% water saturation (Figure 5.28). For each of these three variants, we produce a gather at the sand layer (a) at 1.25 km and (b) 3.25 km depth.

The results displayed in Figures 5.26, 5.27, and 5.28 indicate that compaction does affect the seismic response and, hence, has to be taken into account when constructing a pseudo-well. For wet sand, a strong AVO Class I at shallow depth (about 1.25 km) becomes much weaker as the sediment compacts with depth (about 3.25 km). An AVO Class II at shallow oil sand becomes a weak AVO Class I. A strong AVO Class III at shallow gas sand becomes much weaker with depth.

6 Pseudo-wells: statistics-based generation

6.1 Introduction

Well log data provide information about petrophysical and elastic properties of the subsurface but do not necessarily cover all possible scenarios of interest that we could encounter away from well control. To account for plausible variations in the subsurface, we may decide, for example, to stretch a shale interval or increase or reduce the clay content in the sand. One way of implementing such perturbations is to use statistical simulations. Such simulations should be based on realistic spatial distributions of a single property (e.g., porosity) as well as account for deterministic or statistical relations between two or more properties (e.g., between clay content and porosity).

Assume, for example, that we wish to simulate different geological scenarios for a clastic reservoir by changing the porosity values. Also assume that the probability distribution of porosity is known, and then sample from this distribution a porosity value at each point in depth. At two adjacent locations in the borehole, if we sample independently (i.e., we draw a random sample for the first location and then the second random sample for the second location independently of the value we drew at the first location), we may obtain a very high value of porosity at the first location and a very low value at the second location, or vice versa. Such independent random sampling ignores the spatial continuity expected in porosity variations according to depositional and sedimentological laws. Similarly, if we want to simulate clay content, we cannot simulate it independently from the previously obtained values of porosity, since porosity variations can depend on mineralogical variations in sand and clay content. Hence, proposed rock properties should be correlated in space and also with other properties.

Here, we review basic statistical tools to simulate correlated rock properties according to a spatial continuity model. In the first part we review a well-known method, the Monte Carlo simulation method, which is the most common statistical tool to generate random samples of a given property. We then extend this method by introducing a spatial statistics model to mimic the behavior of rock properties as a function of depth. This allows us to create pseudo-well logs to investigate the elastic signatures and seismic response of different geological scenarios not accounted for in the well.

In the last part of the chapter, we review a statistical method to generate pseudo-logs of petrophysical facies.

6.2   Monte Carlo simulation

Assume that we have a set of well measurements that may include porosity, clay content, water saturation, and the elastic properties. Our goal is to perturb these properties in order to describe geological scenarios not represented in the well. Arguably, the most common technique for achieving this goal is the Monte Carlo simulation. The method can be divided into four steps: (1) assume a geologically plausible probability distribution of a target rock property (e.g., porosity) that covers ranges not present in the well; (2) randomly sample the property from this distribution; (3) use a deterministic transform between rock properties and its elastic attributes, which could be a theoretical rock physics model or simply a relevant empirical trend, and compute the elastic properties for each sampled rock property value; and (4) use this set of computed elastic properties to create a probability distribution of, for example, V_p, V_s, and density.

This method has several applications in geophysics. For example, by generating synthetic seismic data using these distributions, we can assess the uncertainty of seismic attributes, such as intercept and gradient, for rock property (e.g., porosity) ranges not present in the well. If a certain combination of these attributes is present in real seismic data, we can trace it back to the probability of the underlying rock property (Avseth *et al.*, 2005). This statistical approach that relies on a rock physics model is called statistical rock physics. It is used to investigate the effect of porosity, fluid, and lithology on the elastic properties and seismic attributes in a statistical sense.

For example, if we are interested in studying the effect of fluid on the *P*-wave velocity, we can generate a set of samples with different saturations, apply Gassmann's (1951) equation, and study the variations in the computed *P*-wave velocity (Mukerji *et al.*, 2001). As a result, we arrive at a probability distribution of velocity versus saturation. Similarly, we could study the effect of porosity variations, by generating a set of samples with different porosities, applying a rock physics model, such as Raymer's equation, and studying *P*-wave velocity changes. This methodology is easy to apply: if we assume a uniform or Gaussian distribution for the rock properties we want to perturb, we can then estimate the distribution parameters (mean and standard deviation) from the corresponding well log, draw random samples from this distribution, and apply the deterministic rock physics model to obtain the probability distribution of the elastic properties.

This method is instrumental in investigating the elastic response for different geological scenarios not covered by the well. However, this method is not entirely suitable for seismic interpretation. As a matter of fact, because the seismic response depends on the elastic contrast between adjacent layers, we cannot simply randomly sample from the input distribution. If we do this and build a pseudo-well from the simulated samples, we can obtain an unreasonable geological sequence: for example, a sample with very high porosity could be followed by a sample with very low porosity and again by another sample with very high porosity. In other words, when we perform a random simulation from a given distribution we are not assuming any spatial continuity model and each drawn sample is independent from the previously simulated samples. If we apply a rock physics model to such a porosity curve, we may end up with drastic spatial variations in the elastic properties, and, hence, the resulting seismograms will not be geologically realistic.

Therefore, to create realistic pseudo-well logs we must introduce a spatial statistics model which allows us to mimic the vertical continuity of rock and elastic properties in the subsurface. The spatial continuity can be described by a variogram model (Deutsch and Journel, 1996). In the following section, we describe a procedure to create pseudo-logs of porosity that honor the vertical trend and the vertical continuity present in the input well data.

6.3 Monte Carlo simulation with spatial correlation

To create realistic well log data we must account for the spatial correlation present in the well. Spatial correlation is a key concept in geostatistics, since it is the property that allows us to relate a measurement at one point in the reservoir to a measurement of the same property at another point in the same reservoir. A spatial correlation function is a function that describes the correlation between different measurements of the same property as a function of distance. For simplicity in this section we will only refer to one-dimensional problems but the concept can be extended to 2D and 3D.

Figure 6.1 displays porosity and clay content curves in a well (Well A). Let us initially focus on a single property, porosity, in the entire depth interval (Figure 6.2). If we examine one point in the well log, for instance the first sample at the top, it is likely that the points close to this sample have similar values of porosity due to geological continuity. At the same time, points far from it are likely to belong to different layers and/or lithologies and the porosity values there could be very different from the value at the first point in depth. The *vertical correlation function* allows us to mathematically represent these similarities and dissimilarities.

Let us, for example, consider all the pairs of samples with a distance of 1.5 m from each other, read off the two porosity values for each of these pairs and plot the porosity at the second point versus that at the first point (which is 1.5 m higher than the second point) for the entire well (Figure 6.2, top right).

Since the distance between the samples in a pair is small (1.5 m), the two porosity values are likely to be close (except in the proximity to an abrupt lithology change). As a result, the spread of these points will be aligned along the bisector line (diagonal) and the correlation coefficient will be close to one (0.87 in this case).

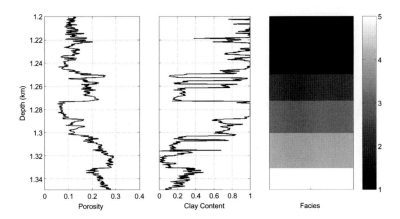

Figure 6.1 Well log data and classification of the depth interval. From left to right, porosity, clay content, and facies into which the interval is broken to ad hoc delineate different classes of porosity and lithology according to their behaviors.

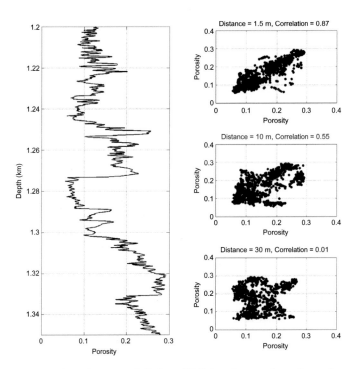

Figure 6.2 Porosity curve (left) from Well A and cross-plots of porosity pairs separated by gradually increasing distance, 1.5, 10, and 30 m, as marked on the plots on the right.

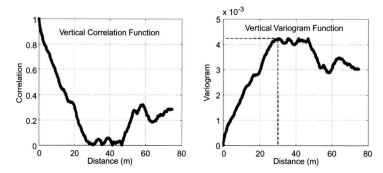

Figure 6.3 Example of experimental vertical correlation function (left) and corresponding experimental vertical variogram function (right). The distance at which the variogram reaches the variance of the data (dashed line) is called the correlation length. It is about 30 m in this example.

If we now gradually increase the distance between the samples in each pair (from 1.5 to 10 to 30 m), the spread of the resulting porosity points around the bisector becomes larger (Figure 6.2, middle and bottom right) and the correlation coefficient severely deteriorates (0.55 and 0.01, respectively).

Conversely, if we reduce the distance between the samples, the spread of points will be closely aligned to the bisector and the correlation coefficient will approach 1. It will become exactly 1 in the extreme case where the distance between the pair of points is zero, and, hence, both points in a pair have the same value.

If we now repeat this sampling procedure for all the possible distances, h, in the well, compute the correlation coefficient for the spread of points for each of these distances, and plot the correlation coefficient as a function of the distance between the points in a pair (Figure 6.3, left), we obtain the *experimental vertical correlation function, $\rho(h)$*, of the data under examination.

Usually, geostatistics deals with *variograms* rather than correlation functions. A variogram is obtained by subtracting the correlation function from 1 and then multiplying the result by the variance of the data (Figure 6.3, right):

$$\gamma(h) = \sigma^2[1 - \rho(h)], \tag{6.1}$$

where σ^2 is the variance (the square of the standard deviation) of the entire dataset.

Generally, we do not expect to observe any correlation in the data when we reach a certain distance, unless there is periodicity in the data or there are systematic errors in the measurements. In other words, after a certain distance the correlation function approaches zero and remains such, and the variogram reaches the variance of the property. The vertical correlation function in Figure 6.3 rapidly approaches zero as the distance between the pair of points increases. This distance is called the *correlation length* and is about 30 m in this example. Notice that, past this distance, the correlation function increases again. This is the result of the length of the well log being finite: we simply run out of samples as we further increase the distance between the pair of

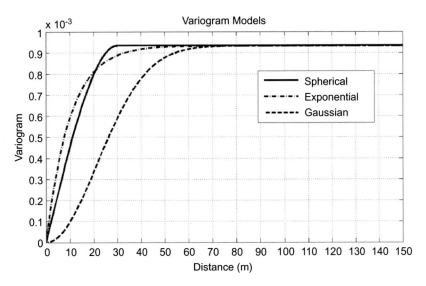

Figure 6.4 Examples of variogram models: Gaussian (dashed line), spherical (solid line), and exponential (dash-dotted line).

points. This is why the tail of the correlation function does not carry useful information and should be ignored.

As shown in Figure 6.3, the variogram estimated from the data can be noisy due to measurement errors, a limited number of samples (especially for large distances), and small- and large-scale heterogeneities. For this reason we usually fit an *analytical variogram* function (model) to the experimental variogram and use this model instead of the experimental one. The three most common variogram models are: Gaussian, spherical, and exponential (Figure 6.4). The Gaussian model generally provides a smoother analytical variogram compared to the spherical and the exponential models. This is because for short distances, this function's increase is gentler than for the other two functions.

A variogram provides useful information about the spatial continuity and variance of the data but it does not provide any information about the absolute values of the property (e.g., porosity). From Figure 6.3 we can infer that the correlation length of our dataset is about 30 meters, and the variance is about 4×10^{-3}, but we cannot infer anything about the mean value. As a result, we cannot use a variogram to directly generate a pseudo-log of the property that we wish to simulate. Instead, we can use it to simulate the residual spatial distribution to be added to the local mean trend of the property.

A procedure for creating a synthetic realization with the spatial features present in the well and using the vertical variogram of the property we simulate is: (1) filter (smooth) the selected rock property (e.g., porosity versus depth) by using a moving average to obtain its depth trend, **m**, from the actual log (any kind of finite impulse response filter could be used for smoothing); (2) build the variogram (as explained in the text) using the difference between the original and smoothed curves and, thus, obtain the data variance and the correlation length; (3) generate a pseudo-log, **w**, of

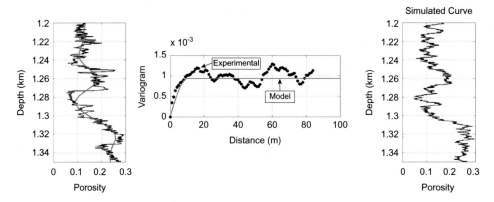

Figure 6.5 Monte Carlo simulation with vertical correlation. Left: Porosity profile from Figure 6.1 (black) and a smoothed (low-frequency) trend (gray). Middle: Experimental vertical variogram of the residuals (black symbols) with the spherical variogram model curve superimposed (gray). Right: Simulated porosity curve.

residual values with zero mean using variogram-based simulations (see below); and, finally, (4) add the residual pseudo-log, **w**, to the smooth trend of the mean, **m**.

To implement this procedure, we first estimate a vertical variogram model, $\gamma(h)$, and create the corresponding 1D spatial symmetrical covariance matrix:

$$\mathbf{C} = \begin{bmatrix} \gamma(0) & \cdots & \gamma(d_{max}) \\ \vdots & \ddots & \vdots \\ \gamma(d_{max}) & \cdots & \gamma(0) \end{bmatrix}, \tag{6.2}$$

where d_{max} is the maximum distance.

Next, the Cholesky decomposition (e.g., Tarantola, 2005) of this matrix is computed:

$$\mathbf{C} = \mathbf{R}\mathbf{R}^T \tag{6.3}$$

where **R** is the lower triangular matrix and $\mathbf{R}^T$ is the transpose of **R**.

To simulate a random vertically correlated vector, **w**, we first generate a random uncorrelated vector, **u**, of the same length as the initial data vector, normally distributed with zero mean and standard deviation equal to 1 ($\mathbf{u} \sim N(0,1)$). Next, we multiply this vector, **u**, by matrix, **R**, ($\mathbf{w} = \mathbf{R}\mathbf{u}$) and finally add to it the background trend containing locally varying mean, **m**:

$$\mathbf{v} = \mathbf{m} + \mathbf{w} = \mathbf{m} + \mathbf{R}\mathbf{u}. \tag{6.4}$$

The randomness present in vector, **u**, is the source of randomness in the final realization, **v**.

An example of this operation is shown in Figure 6.5. On the left is the actual porosity curve with a background trend obtained by smoothing this curve. The residual curve

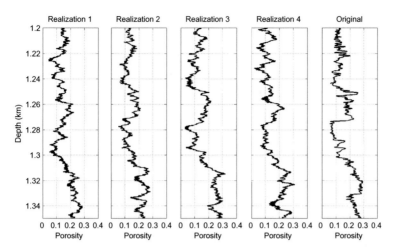

Figure 6.6 Left to right: Four random realizations of porosity and the original porosity curve, as marked in the plots.

is computed as the difference between the actual porosity curve and the background trend. The experimental variogram of this residual is shown in the middle plot. This experimental variogram is approximated by a *spherical analytical variogram* function. Finally, a porosity pseudo-curve is simulated as explained in the text (Figure 6.5, right). The same technique is used to produce four different random realizations of the porosity curve (Figure 6.6).

6.4 Monte Carlo simulation within facies

To avoid mixing drastically different lithologies present in the well, it is instrumental to subdivide the interval under examination into several separate ad hoc facies intervals (do not confuse with petrophysical lithofacies) and apply the above-described Monte Carlo simulation with spatial correlation separately within each of these facies. The procedure is the same as described in the previous section but now it is applied independently to each facies interval of the well. An example of such subdivision is shown in Figure 6.1 where the entire interval is broken into five facies based on the porosity and clay content curves. This subdivision is especially important where we wish to preserve the natural contrast in rock properties present in the well, such as between the sand interval situated approximately between the depths of 1.25 and 1.27 km in Well A and the underlying shale (Figure 6.1).

In order to follow the main geological features of the subsurface, the simulation procedure should be applied separately in different intervals with locally varying trends and interval-dependent variograms. In Figure 6.7 we show an example where we apply such regional Monte Carlo simulation with spatial correlation in each interval. The

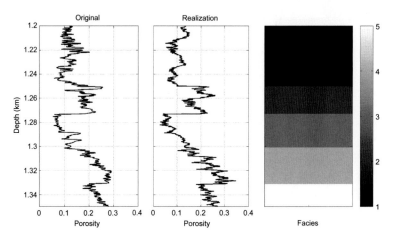

Figure 6.7 Regional Monte Carlo simulation with vertical correlation: actual porosity log (left), a realization of simulated porosity (middle), and interval classification (right).

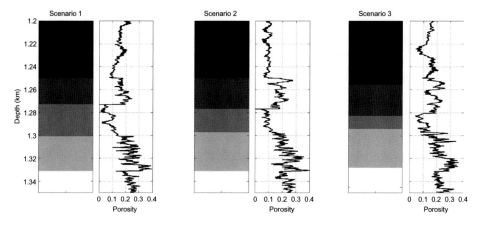

Figure 6.8 Three realizations of porosity with modified interval thickness (mid-layer thickness decreases from left to right).

advantage of the regional approach compared to the global one is that we can reproduce local heterogeneities and preserve strong contrasts between different layers. Moreover we can also vary the thickness of the layers, and by so doing build the corresponding pseudo-logs where the facies are stretched or shortened (as shown in Figure 6.8, where we gradually reduce the thickness of the middle interval with low porosity).

6.5 Stochastic simulation of related variables

So far we have only considered one single variable (porosity). However, certain rock properties should not be simulated independently of each other since they may be related (e.g., clay content and porosity). A procedure similar to that described in the previous sections can be used to simultaneously simulate two or more different curves, for example porosity and clay content, at the same time, including vertical correlation and the correlation between the properties. The method is illustrated for two variables, but it can be extended to any finite number of variables. As in the procedure described for a single variable, we compute the local mean (smoothed) trends, $\mathbf{m}_1$ and $\mathbf{m}_2$, for both variables, fit each of them with the *same type* of analytical variogram model, and estimate the correlation lengths of the two original curves. Bear in mind that if these two properties are correlated, their respective correlation lengths should be close to each other. Hence, in the following step, we select a *single correlation length* for both variables, which could be either of the two or simply their mean. This operation provides us with a single analytical variogram for both variables but with different variances. We normalize this common analytical variogram with the variance equal to 1. The corresponding covariance matrix of this common analytical variogram is $\mathbf{C}$.

Next we form the covariance matrix, $\mathbf{S}$, for these two variables which is a 2 by 2 symmetrical matrix. If the first variable is porosity and the second variable is clay content, this matrix is:

$$\mathbf{S} = \begin{bmatrix} \sigma_\phi^2 & \sigma_{\phi,C} \\ \sigma_{\phi,C} & \sigma_C^2 \end{bmatrix}, \tag{6.5}$$

where, as usual, the variances σ_ϕ^2 for porosity and σ_C^2 for clay content are defined as

$$\sigma_\phi^2 = \sum_{i=1}^{n} (\phi_i - \mu_\phi)^2, \quad \sigma_C^2 = \sum_{i=1}^{n} (C_i - \mu_C)^2, \tag{6.6}$$

n is the number of the samples and μ_ϕ and μ_C are the means of the respective variables. Also $\sigma_{\phi,C}$ is the covariance of the porosity and clay content:

$$\sigma_{\phi,C} = \sum_{i=1}^{n} (\phi_i - \mu_\phi)(C_i - \mu_C). \tag{6.7}$$

We then compute the Kronecker product

$$\mathbf{K} = \mathbf{S} \otimes \mathbf{C} \tag{6.8}$$

and apply the Cholesky decomposition, $\mathbf{K} = \mathbf{R}\mathbf{R}^T$. Both $\mathbf{K}$ and $\mathbf{R}$ will then have size $2n$ by $2n$.

In order to simulate two correlated random vectors with n samples, we generate a random vector, $\mathbf{u}$, with $2n$ samples normally distributed with zero mean and standard deviation 1 ($\mathbf{u} \sim N(0,1)$), multiply this vector by matrix, $\mathbf{R}$, and, finally, add to it the background trend containing locally varying means:

$$\begin{bmatrix} \mathbf{v}_1 \\ \mathbf{v}_2 \end{bmatrix} = \begin{bmatrix} \mathbf{m}_1 \\ \mathbf{m}_2 \end{bmatrix} + \mathbf{R}\mathbf{u} \tag{6.9}$$

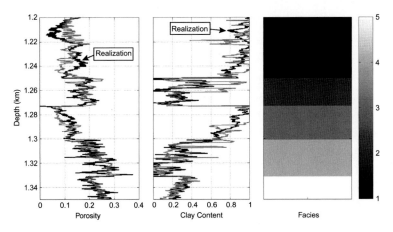

Figure 6.9 Simultaneous facies-based Monte Carlo simulation with vertical correlation of porosity and clay content: porosity realization (left), clay content realization (middle), and interval classification (right). The original curves are gray while the realizations are shown in black.

Figure 6.9 shows an example of simultaneous co-simulation of porosity and clay content using regional Monte Carlo simulation with five facies intervals.

6.6   Examples and sensitivity analysis

In the example shown in Figure 6.10, we preserve the interval classification shown in Figure 6.1 and perform 50 different simulations of porosity and clay content by honoring the correlation between the two properties, and the depth trends of each property. The results indicate that a wide variety of geologically plausible outcomes can be generated from deterministic well data.

Several parameters can be varied to further expand the outcome space. One of them is the correlation length that can be artificially changed once the analytical variogram model is found based on the experimental variogram. The results of this exercise are shown in Figure 6.11. We observe that by reducing the correlation length (e.g., to 10 from the original 30 m) we increase the spatial frequency of the outcome; that is, the realization becomes "noisier" (Figure 6.11, left) since this operation is equivalent to forcibly reducing the spatial continuity of the original data. In contrast, if we artificially increase the correlation length (e.g., to 50 from the original 30 m), the resulting realizations become more continuous (Figure 6.11, right).

To expand the perturbation range of measured variables, the randomly generated residuals can be added to any geologically plausible low-frequency background trend. Such a trend can be obtained directly at the well by smoothing the respective curves. Once this trend is computed, it can be manually perturbed to increase or reduce the values within a facies. Moreover, we can literally move the well through geologic time

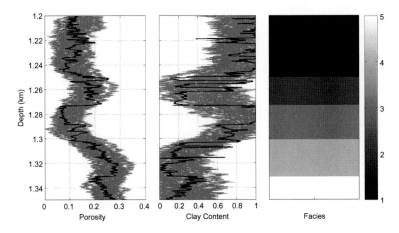

Figure 6.10 Multiple realizations (gray) of porosity (left) and clay content (middle) honoring the depth trend and the vertical correlation observed at the well location. The actual well data curves are superimposed in black. The subdivision of the interval into facies (same as in Figure 6.1) is shown in the right-hand plot.

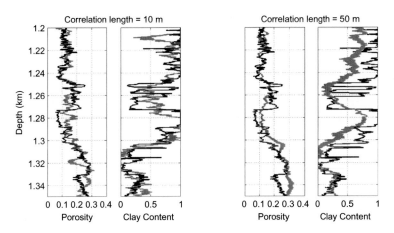

Figure 6.11 Sensitivity analysis of a simulation to the correlation length as marked in the plots. The actual well logs are superimposed in black.

by changing its depth interval. For this purpose, compaction (or depth) trends can be used, such as discussed in Chapter 5. Here we provide an example using porosity (ϕ) versus depth (z) reduction equations by Ramm and Bjørlykke (1994):

$$\phi_{sh}(z) = \phi_{sh}^0 e^{-\alpha(z-z_0)};$$
$$\phi_{ss}(z) = \begin{cases} \phi_{ss}^0 e^{-\beta(z-z_0)}, & z \leq z^c, \\ \phi_{ss}(z^c) - \gamma(z-z^c), & z > z^c, \end{cases} \tag{6.10}$$

where the subscripts sh and ss are for shale and sand, respectively; z_0 is a reference depth; ϕ_{sh}^0 and ϕ_{ss}^0 are the shale and sand porosity, respectively, at z_0; z^c is the depth of

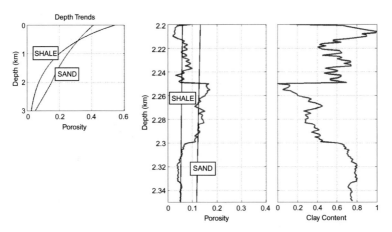

Figure 6.12 Porosity and clay content simulated in the depth interval from 2.20 to 2.35 km using the compaction depth trends from Eq. (6.10) and the statistics obtained at the actual well location between 1.20 and 1.35 km. In the porosity track, the two monotonic curves are the porosity trends for sand and shale, as marked.

the onset of diagenetic cementation in sand; and α, β, and γ are empirically calibrated constants whose units are km^{-1} (see also Section 5.3).

In the following example, we use $z_0 = 1.00$ km; $\phi_{sh}^0 = 0.20$; $\phi_{ss}^0 = 0.25$; $z^c = 2.00$ km; $\alpha = 1.00$ km^{-1}; $\beta = 0.50$ km^{-1}; and $\gamma = 0.10$ km^{-1}. We move the well 1 km down from its current depth, so that now the depth interval starts at 2.2 km. We use Eqs (6.10) to simulate the porosity reduction with depth. The clay content is then co-simulated according to the correlation observed at the well location. The resulting porosity and clay content curves are shown in Figure 6.12.

6.7   Pseudo-logs of facies*

The above-described methodology helps generate pseudo-logs of continuous rock properties, for example, porosity and/or clay content. Sometimes we have to deal with *discrete variables*, such as lithofacies or ad hoc defined facies. These could be just shale and sand. Additional features, such as the shale content and pore fluid, can be added to the description of the intervals, which will further increase the number of categories (facies). For example, such facies may include shale, sandy shale, shaly sand, clean oil sand, clean gas sand, and clean wet sand, six categories altogether. Facies can be *numbered* by assigning an integer number to each category. As a result, we will be dealing with Facies 1, Facies 2, …, Facies N.

---

* This part was modified from work originally published by SEG (Grana *et al.*, 2012)

The goal now is to generate geologically plausible layered sequences of these facies. A common technique used in geostatistics for generating a random layered sequence is the Markov chain Monte Carlo simulation, usually abbreviated as McMC (Grana et al., 2012). This technique can be tailored to simulate facies sequences that capture realistic features in geology and deposition (Krumbein and Dacey, 1969).

Markov chains are based on a set of conditional probabilities that describe the dependency of the facies occurring at a given location with other facies located above (an upward chain) or below (a downward chain). A chain is called a first-order chain if the transition from one facies to another depends only on the immediately preceding facies. The conditional probability of the transitions are the elements of the so-called transition matrix, $\mathbf{P}$, where the element $\mathbf{P}_{ij}$ is the probability of a transition from facies i to facies j located immediately below facies i.

Consider a set of three facies: sand, silty sand, and shale. An example of the transition matrix is

$$\mathbf{P} = \begin{array}{ccc} \text{sh} & \text{si} & \text{sa} \\ \begin{bmatrix} 0.90 & 0.05 & 0.05 \\ 0.00 & 0.95 & 0.05 \\ 0.05 & 0.00 & 0.95 \end{bmatrix} & \begin{array}{l} \text{sh} \\ \text{si} \\ \text{sa} \end{array} \end{array}, \tag{6.11}$$

where abbreviations sh, si, and sa indicate shale, silty sand, and sand, respectively.

In this matrix, the rows correspond to the probabilities for shale, silty sand, and sand at a given depth interval while the columns correspond to the probabilities for shale, silty sand, and sand in the next (deeper) layer. Specifically, in this matrix, the probability of finding shale on top of shale is 90%, the probability of finding shale on top of silty sand is 0%, and the probability of finding shale on top of sand is 5%. This means that in this example we can never have shale situated on top of silty sand or silty sand on top of sand. The terms on the diagonal of the transition matrix are related to the thickness of the layers: the higher the numbers on the diagonal, the higher the probability to observe no transition (i.e., high probability that a facies has a transition to itself). This high probability translates into the increased thickness of a given layer.

In the following example, a clastic sequence including a reservoir filled with oil (10% water saturation) starts at 2.00 km. The oil/water contact is located at 2.10 km. The reservoir is capped by shale.

We model a facies sequence in the well by a first-order Markov chain and then distribute the corresponding rock properties accordingly. The first sample in the profile is shale. In the next step, the facies number i is sampled from the conditional probability $P(F_i \mid F_{i-1})$, where F_i stands for ith facies. This process is continued until the bottom of the interval is reached. This specific realization is the desired facies profile (Figure 6.13).

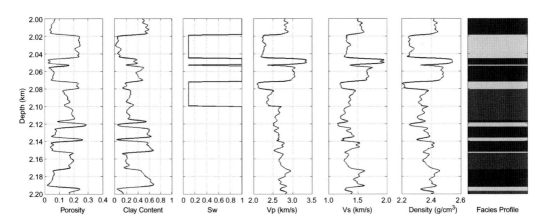

Figure 6.13 Synthetic well log generated as explained in this section. Left to right: Porosity, clay content, water saturation, the *P*- and *S*-wave velocity, bulk density, and facies profile (black is shale, dark gray is silty sand, and light gray is sand). After Grana *et al.* (2012).

Next, within each facies, we generate pseudo-logs of porosity and clay content. As explained before, these two pseudo-logs must be vertically correlated within each facies and, at the same time, correlated with each other to honor the often seen interdependence of porosity and clay content.

These pseudo-logs of porosity and clay content are generated by sampling from three bivariate Gaussian distributions, one for each facies. Together, these three distributions form a single bivariate multimodal distribution (Figure 6.14). These three distributions can be estimated from well log data from the existing wells in the field, from nearby fields, or assumed from geological prior knowledge of the field. From these Gaussian distributions, for each facies we now have the mean values of porosity and clay content, their variances, their covariances, and correlations. The mean values are used to build the low-frequency trends, $\mathbf{m}_1$ and $\mathbf{m}_2$. Since the facies thickness is small in this example, we used a constant trend in each facies. Variances and covariances (Eqs (6.6) and (6.7)) are used to build the matrices, $\mathbf{S}$, (Eq. (6.5)), one for each facies.

For each facies we assume a Gaussian analytical variogram model with correlation lengths 3, 10, and 4 m, for shale, silty sand, and sand, respectively (Figure 6.15). These variograms have three corresponding matrices, $\mathbf{C}$, one for each facies.

We multiply (Kronecker product) the covariance matrices, $\mathbf{S}$, of each distribution by the covariance matrices, $\mathbf{C}$, obtained from the variograms (Eq. (6.8)) and take the Cholesky decomposition. The result, $\mathbf{R}$, is multiplied by a normally distributed random vector and added to the local trends (Eq. (6.9)). We repeat this procedure for each facies and obtain, as a result, three pseudo-logs of porosity and clay content, one for each facies. Finally, the clay content and porosity pseudo-curves thus simulated are reassembled into the complete vertical profiles according to the facies profile simulated earlier.

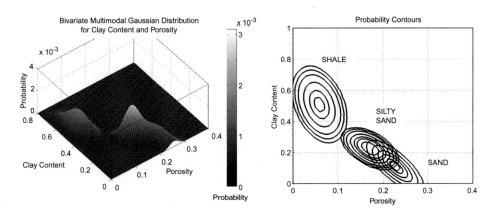

Figure 6.14 Bivariate multimodal Gaussian distribution for the clay content and porosity (left) and the corresponding probability contours (right). After Grana *et al.* (2012).

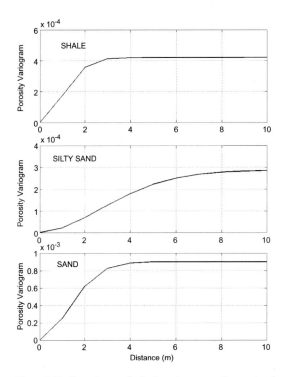

Figure 6.15 Gaussian analytical variograms of porosity for shale, silty sand, and sand (top to bottom). After Grana *et al.* (2012).

Once these inputs are generated, they can be used in a selected rock physics velocity–porosity model to create the elastic property profiles (the *P*- and *S*-wave velocity, and bulk density). In this specific example, the soft-sand model was used by assuming that the rock contains only two minerals, quartz and clay. Finally, to account for the natural spread of data around an ideal model trend, we added to each of the

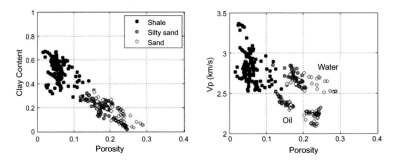

Figure 6.16 Rock physics cross-plots for the stochastically generated rock properties. Left: Clay content versus porosity. Right: The *P*-wave velocity versus porosity. After Grana *et al.* (2012).

modeled curves a random error with a 1 m vertical correlation length. The results are shown in Figure 6.13 on the left of the facies profile. Figure 6.16 shows cross-plots of the rock properties thus simulated in the interval under examination.

6.8   Spatial simulation of rock properties and reflections

Stochastic generation of a 3D volume of rock properties is more complex than the 1D pseudo-log generation. A 3D volume of a rock property is a collection of 1D pseudo-logs regularly spaced over the seismic grid. However, this 3D volume cannot be built by simply simulating a set of pseudo-logs independently of each other, because the property has to be correlated horizontally and this correlation has to be geologically consistent. To achieve this goal, a spatial statistics model is required to describe geological continuity in 3D, similar to that described in the preceding sections for pseudo-log generation where the spatial continuity model was represented by a vertical variogram. Various methods and models can be used in 3D, such as spatial variograms or training images, as described in Deutsch and Journel (1996).

An effective class of such methods is sequential simulations. These are geostatistical techniques that can be used to generate realizations of a probability density function of either discrete or continuous properties. A sequential simulation procedure produces realizations of the desired property by sequentially visiting the grid cells of a 1D, 2D, or 3D space along a random path. In each cell, the simulated value is drawn from the local conditional distribution, which depends on the prior distribution and on the previously simulated values in the neighborhood of the given cell. This procedure is applied to cover all the cells of the grid.

Sequential simulation methods can be divided into two major categories: two-point and multipoint geostatistics. The former algorithms are faster than the latter as they only account for the correlation between two spatial locations at a time, with the spatial continuity of the property distribution being ensured by variogram models.

The two most common algorithms in two-point geostatistics are sequential indicator simulation and sequential Gaussian simulation (Deutsch and Journel, 1996). Sequential indicator simulation deals with discrete random variables (e.g., facies in reservoir modeling), while sequential Gaussian simulation deals with continuous random variables (e.g., porosity). In contrast, multipoint geostatistics takes into account the correlation between multiple spatial points. As a result, it is usually quite difficult to find an analytical model for the associated conditional probability. Both methods allow for including background trends derived from measurements or rock physics models to describe non-stationary behaviors of the properties.

A commonly used method within the two-point sequential simulation approach is probability field simulation where a 3D spatially correlated error, $\varepsilon(x, y, z)$, is rescaled by a locally varying variance, $\mathbf{v}(x, y, z)$, and added to a locally varying mean, $\mathbf{m}(x, y, z)$, as

$$\mathbf{p}(x, y, z) = \mathbf{m}(x, y, z) + \mathbf{v}(x, y, z)\varepsilon(x, y, z). \tag{6.12}$$

This is the method we use in the following example. The volume containing the spatial correlated error can be generated, for example, by using sequential Gaussian simulation, while the volume of the locally varying mean can be derived by interpolating field data or from an analytical trend model.

In this section we show a geostatistical simulation of rock properties (Figure 6.17) where porosity was simulated using a low-frequency trend from a real clastic (quartz/clay) dataset that also included the clay content and water saturation (the hydrocarbon in this example is oil). Based on this porosity simulation, the clay content and water saturation were co-simulated using the relations present in the original data set.

To obtain the properties of the pore fluid, we used the Batzle and Wang (1992) equations where the brine salinity was 80,000 ppm; oil API gravity was 30; gas gravity was 0.70; the gas-to-oil ratio (GOR) was 400; and the pore pressure and temperature were 15 MPa and 80° C, respectively. Then the soft-sand model was applied to produce the elastic properties of the volume. The simulated porosity, clay content, and water saturation are shown in Figures 6.17 to 6.19.

The bulk density as well as V_p and V_s, all computed from the porosity, clay content, and saturation using the soft-sand model, are shown in Figures 6.20 to 6.22. The respective P-wave impedance and Poisson's ratio are shown in Figures 6.23 and 6.24.

The seismic traces were computed at each horizontal station for the respective vertical distribution of the elastic properties. We used a 30-Hz Ricker wavelet and the same ray tracing algorithm we described and used in previous chapters. The resulting normal incidence and far-offset amplitudes are shown in Figures 6.25 and 6.26, respectively. In the latter, the incidence angle was approximately 45°.

In this example, both the impedance and Poisson's ratio in the reservoir are smaller than in the surrounding strata. As a result, in Figures 6.25 and 6.26, we observe a

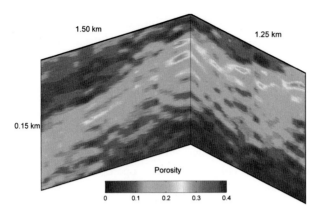

Figure 6.17 Simulated porosity volume (as explained in the text) displayed in two orthogonal cross-sections. The horizontal and vertical distances are marked along the respective directions. For color version, see plates section.

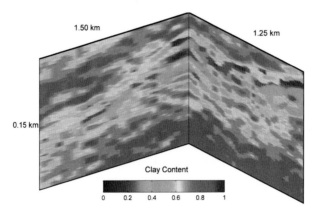

Figure 6.18 Same as Figure 6.17 but for the clay content. For color version, see plates section.

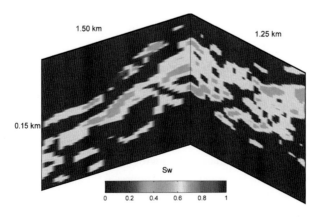

Figure 6.19 Same as Figure 6.17 but for water saturation. For color version, see plates section.

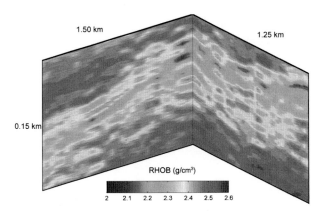

Figure 6.20 Same as Figure 6.17 but for the bulk density. For color version, see plates section.

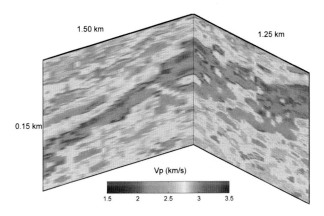

Figure 6.21 Same as Figure 6.17 but for the *P*-wave velocity. For color version, see plates section.

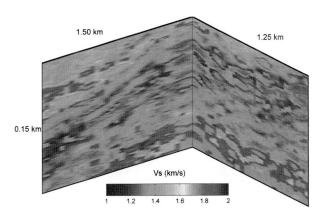

Figure 6.22 Same as Figure 6.17 but for the *S*-wave velocity. For color version, see plates section.

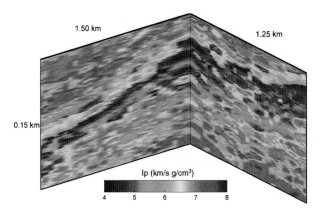

Figure 6.23 Same as Figure 6.17 but for the *P*-wave impedance. For color version, see plates section.

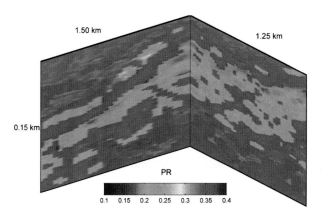

Figure 6.24 Same as Figure 6.17 but for Poisson's ratio. For color version, see plates section.

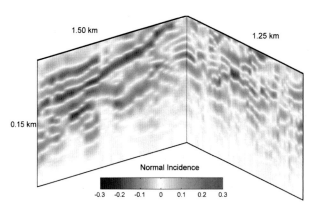

Figure 6.25 Same as Figure 6.17 but for normal incidence amplitude. For color version, see plates section.

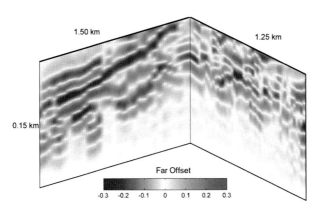

Figure 6.26 Same as Figure 6.17 but for far-offset amplitude. For color version, see plates section.

Class III AVO response with a normal-incidence negative amplitude becoming increasingly negative with increasing offset.

It is important to mention in conclusion that in the last few decades, a number of methodologies have been introduced to simulate rock properties in the subsurface in order to describe geologically consistent scenarios. The least computationally intensive methods are generally easy to use but they do not capture the full spectrum of geological variability and the respective spatial heterogeneity. In contrast, the most computationally intensive methods aimed at a more comprehensive representation of geology (e.g., object models and process models) are difficult to manipulate to match multiple observations. More specifically, in the context of such models, it is fairly straightforward to create a geo-object such as a channel based on assumed properties including length, tortuosity, width, and depth and then place this object in the virtual subsurface. The most complicated part comes when this object (channel) has to conform to factual observations in multiple wells at concrete geographical locations.

This is a historically common tradeoff in science that attempts to model nature. A model can be as complex and comprehensive as desired to approximate natural occurrences, but in return it has to be changed every time a new datum becomes available. Hence, a comprehensive model is usually not predictive and a simple model is not entirely comprehensive. Bearing this in mind, in this book we follow the principle that a model has to be "as simple as possible but not simpler."

Part III

From well data and geology to earth models and reflections

7 Clastic sequences: diagnostics and V_s prediction

7.1 Unconsolidated gas sand

Figure 7.1 shows data from a well penetrating a sand/shale sequence in a gas-bearing offshore well. The sand interval includes two sub-intervals. The upper one contains gas while the lower one is wet. Only the P-wave velocity is available in this set of well data. V_p-only fluid substitution (Eq. (2.15)) was conducted and the resulting wet-rock density, P-wave velocity, and impedance were computed (Figure 7.1, bottom).

In order to conduct the rock physics diagnostics in the interval under examination, we need to bring the entire interval to the common fluid denominator, which, in this case is the formation brine. The impedance versus porosity plots, color-coded by GR for both the in situ and wet conditions are shown in Figure 7.2.

A theoretical model seemingly appropriate for this dataset is the soft-sand model (Eqs (2.29) to (2.31)) where the differential pressure is 50 MPa, the coordination number is 7, and the shear modulus correction factor is 1. The elastic properties of quartz and clay are those listed in Table 2.1, except for the clay density which we assume is 2.65 g/cm^3 in this example, the same as for quartz. Also, here, the bulk modulus of water is 2.88 GPa and its density is 1.03 g/cm^3. The respective values for gas are 0.13 GPa and 0.26 g/cm^3. The resulting model curves superimposed on the wet-rock impedance versus porosity cross-plot are shown in Figure 7.2 (right).

The diagnostics shown in Figure 7.2 indicate that the soft-sand model explains the observed dependence of the P-wave elastic properties on porosity and clay content, the latter represented by GR. However, this result does not guarantee that the model matches the data at *each* depth station. To confirm this match, we need to have the clay content (C) along the interval.

A common way of estimating the clay content is to linearly transform the GR into C between selected pure-quartz and pure-shale points. The former can be the minimum GR value (GR$_{min}$) while the latter can be the maximum GR value (GR$_{max}$). The respective clay-content equation is

$$C = \frac{\text{GR} - \text{GR}_{min}}{\text{GR}_{max} - \text{GR}_{min}}. \tag{7.1}$$

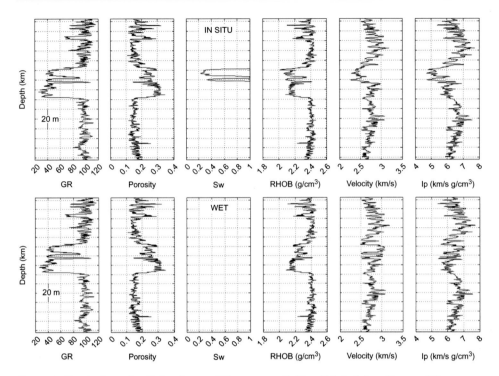

Figure 7.1 Depth plots of well data for an offshore gas well. Top: Original data. Bottom: Wet conditions with the density, velocity, and impedance computed by fluid substitution.

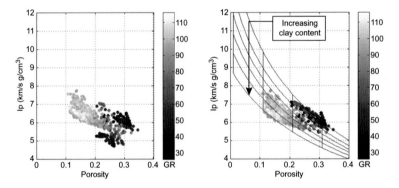

Figure 7.2 The *P*-wave impedance versus porosity for the interval shown in Figure 7.1, color-coded by GR. Left: In situ conditions. Right: Wet-rock impedance computed from the in situ elastic properties by means of fluid substitution. The model curves on the right are from the soft-sand model as explained in the text. The arrow indicates increasing clay content for the model curves.

Of course, this method has its limitations and can even be misleading if traces of radio-active elements are present in the interval that affect the natural radioactivity of rock.

In Eq. (7.1), GR_{min} and GR_{max} values can be assigned manually and they do not have to necessarily be the exact extrema found in the GR curve. Here we select $GR_{min} = 35$ and $GR_{max} = 105$. Of course, with these numbers, Eq. (7.1) will produce a negative clay

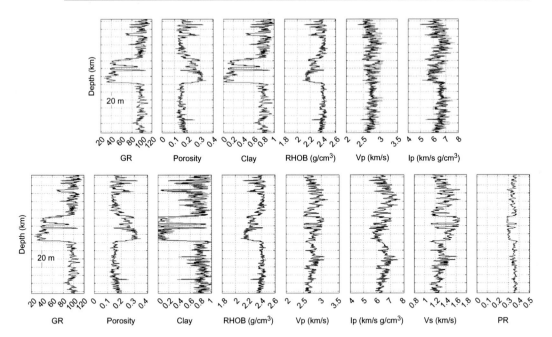

Figure 7.3 Depth plots of well data for the example in the text. All elastic properties are for the wet-rock conditions. Top: GR, porosity, clay content computed from GR as explained in the text, the wet-rock density, velocity, and impedance computed from the soft-sand model using this clay content. Bold gray curves in the density, velocity, and impedance tracks are for the measured values, while the superimposed black curves are model-based. Bottom: Same as top except for the clay content recomputed using the measured velocity (black) on top of the GR-based clay content estimate and the model-based density, velocity, and impedance computed now from the updated clay content. The well data are shown by bold gray curves while the model-based results are black. The last two tracks show the model-based S-wave velocity and Poisson's ratio.

content $(C < 0)$ where GR < 35 and $C > 1$ where GR > 105. To correct for this artifact, we simply assign $C = 0$ where $C < 0$ and $C = 1$ where $C > 1$.

The resulting clay content is plotted versus depth in Figure 7.3 (top). Using this clay content as well as the total porosity, we can now compute the model-based density, velocity, and impedance in the well for wet conditions.

The resulting elastic properties match the trends of the measured profiles but fail to exactly reproduce them. However, in order to predict V_s using the soft-sand model, we do need the clay content curve that produces an accurate match between V_p measured and V_p predicted. This curve can be computed by running the soft-sand model in reverse to find C from the measured porosity and V_p. The resulting C curve is plotted in Figure 7.3 (bottom) and appears to qualitatively follow the clay content curve computed from GR with certain quantitative deviations. The wet-rock V_p and I_p recomputed with this updated clay content accurately match the data (Figure 7.3, bottom) except for a couple of depth points in the middle of the interval where the deviation between the black and gray curves is discernible. The reason for this occasional deviation is that the measured velocity is higher than that computed using the soft-sand model with the

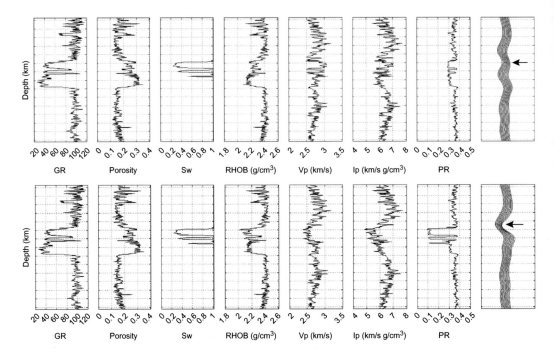

Figure 7.4 Depth plots of well data for the example in the text. Top: Wet-rock conditions. Bottom: In situ conditions. The elastic properties are model-based. The synthetic seismic gathers at the well are produced by using a 30 Hz Ricker wavelet. The maximum incidence angle is about 45°. The arrows point to the features in the seismic response. The saturation curve in the upper graph is for the in situ conditions. The saturation curve in the bottom plot is also the original in situ curve.

measured porosity and updated clay content. This may be due to small inconsistencies in the measured values or a failure of our assumption that the rock contains only two minerals, clay and quartz, and obeys the soft-sand model velocity–porosity transform. However, such exceptions are few, which allows us as to accept the modeling as appropriate for this case study.

This result allows us to compute V_s using the same rock physics model applied now to the porosity and the updated clay content (Figure 7.3). We can also use the model with the same clay content input and compute V_s at wet conditions as well as at in situ saturation (Figure 7.4). Next, by using the predicted V_s, we can compute synthetic seismic traces for these conditions (Figure 7.4).

The response of the wet sand is noticeably different from that of the same sand but filled with gas: in the first case we observe a small peak at the top of the wet sand preceded and followed by troughs. In the second case we observe a well-pronounced trough with a strong Class III AVO signature.

The key in this case is predicting the S-wave velocity that is necessary for generating a synthetic seismic gather. To accomplish this task we established a rock physics model that matched the compressional velocity data. Because this model is a theoretical one,

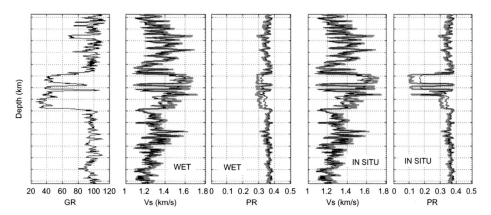

Figure 7.5 Depth plots of GR (left), the *S*-wave velocity, and Poisson's ratio. Bold gray curves are for the two latter properties as predicted by the soft-sand model (Figure 7.4) while the black curves are from GC. The second and third tracks are for wet conditions while the fourth and fifth tracks are for in situ conditions (gas).

it not only produces V_p from porosity, clay content, and fluid bulk modulus and density, but also predicts V_s using the same inputs. A more traditional approach is to use one of the empirical V_s predictors discussed in Chapter 2. When doing so, one has to remember that most of the empirical V_s predictors have been designed for wet rock. This means that if only V_p is available in the well, fluid substitution to wet conditions has to be conducted prior to predicting V_s from V_p. To account for the in situ conditions where hydrocarbons may be present in the well, and because the shear modulus does not depend on the pore fluid, this V_s has to be corrected as

$$V_{sInSitu} = V_{sWet} \sqrt{\frac{\rho_{bWet}}{\rho_{bInSitu}}}, \tag{7.2}$$

where the subscripts are self-explanatory.

Let us now use (one of the most popular) the Greenberg and Castagna (1992) predictor (GC) and apply it to V_{pWet} computed from $V_{pInSitu}$ using V_p-only fluid substitution. The results are shown in Figure 7.5 for both wet and in situ conditions.

Both V_s and Poisson's ratio, as predicted by the soft-sand model and GC, are essentially the same in the shale. However, in the sand, for both wet and in situ conditions, the V_s from the soft-sand model is larger than that predicted by GC. As a result, the Poisson's ratio from the soft-sand model is smaller than that from GC.

The dilemma we are facing is which one of the two predictions to use. Instead of arguing which of the two results is correct, we need to reformulate the question by asking how much the difference between the two V_s predictors affects the synthetic seismic response and, hence, our ability to interpret field seismic data in terms of porosity,

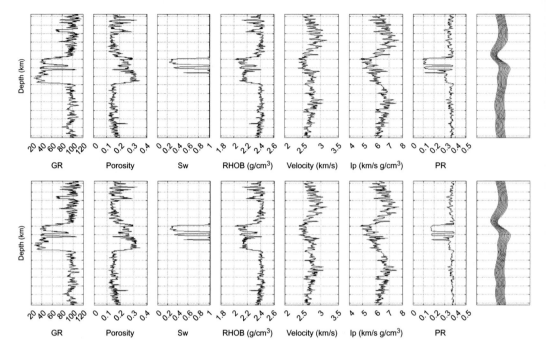

Figure 7.6 Top: Same as the bottom graph in Figure 7.4. Bottom: Same as top but using the Greenberg and Castagna (1992) predictor. The plots are for the in situ (gas) conditions.

lithology, and fluid. The synthetic seismic results shown in Figure 7.6 indicate that the synthetic gathers generated using both predictions are practically the same.

The main lesson learned is that the precision of the answer has to be comeasurable with the goal, which, in this case, is the seismic response. Clearly here the synthetic seismic gather does not perceptibly change between the two V_s predictions. This means that both answers are correct in the context of seismic interpretation.

Let us then adopt the soft-sand-model based prediction and use it to construct a forward-modeling AVO applet as shown in Chapter 4. Based on the well data, we set the porosity ranges in the sand and shale as 0.20 to 0.35 and 0.10 to 0.25, respectively. The clay content range in the sand is zero to 0.20 while that in the shale is 0.60 to 1.00.

Figure 7.7 shows an implementation of the applet to assess reflection between shale and medium-porosity gas sand. Figure 7.8 is for reflection between shale and low-porosity gas sand. Figures 7.9 and 7.10 are for reflection between gas sand and wet sand and between wet sand and shale, respectively.

Notice that the AVO response shown in Figures 7.7 and 7.9 is present in the synthetic seismic gather shown in Figure 7.6 (e.g., strong Class III AVO between shale and gas sand). Because this AVO modeling is conducted at a sharp interface between two infinite half-spaces, it usually exaggerates the response discernible in synthetic or real seismic data. At the same time, by delivering the result in its pure form, not necessarily

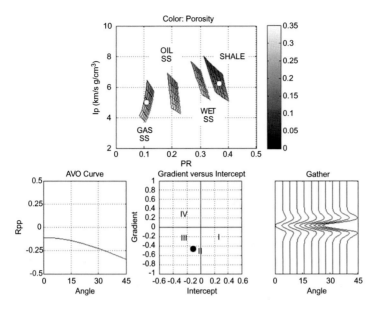

Figure 7.7 AVO modeling applet based on the soft-sand model established by rock physics diagnostics. Reflection between shale and medium-porosity gas sand.

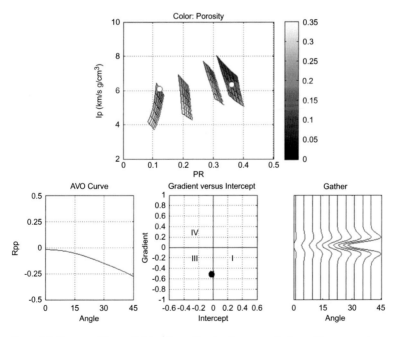

Figure 7.8 Same as Figure 7.7 but for reflection between shale and low-porosity gas sand.

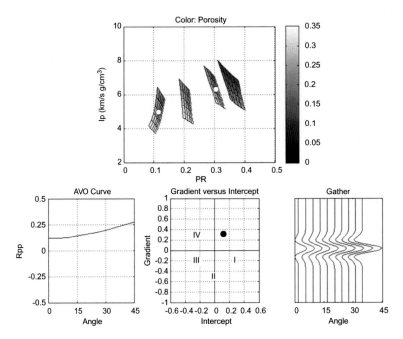

Figure 7.9 Same as Figure 7.7 but for reflection between gas sand and wet sand.

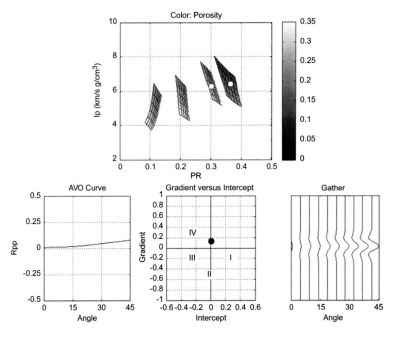

Figure 7.10 Same as Figure 7.7 but for reflection between wet sand and shale.

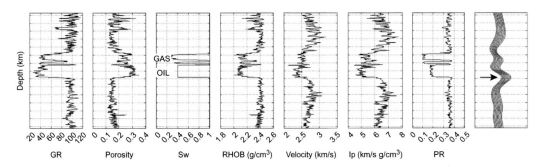

Figure 7.11 Same as Figure 7.6 but for the originally wet sand interval filled with oil as explained in the text. The arrow points at the positive peak at the interface between the oil sand and underlying shale.

present in real data, this modeling guides us towards what to look for in real seismic data for physics-based interpretation.

In the next scenario, let us fill the clean wet sand interval underneath the gas sand with oil. Let us assume that the oil has an API of 30 and a gas-to-oil ratio of 300. These inputs to the Batzle–Wang (1992) fluid property model (Chapter 2) give the bulk modulus of oil as 0.53 GPa and its density as 0.65 g/cm³. Let us also assume water saturation of 40% in the oil sand. With the water bulk modulus of 2.88 GPa and its density of 1.03 g/cm³, the effective bulk modulus of the oil/water mixture is 0.79 GPa and its density is 0.80 g/cm³.

The resulting synthetic gather is shown in Figure 7.11. Because the porosity of the lower sand interval is larger than that of the gas sand, its impedance, if filled with oil, is smaller than that of the overlying gas sand. As a result, we observe two negative peaks, one at the top of the gas sand and the other at the top of the oil sand. The strongest event in this hypothetical situation is a positive peak at the interface between the oil sand and underlying shale, with the positive amplitude gradually increasing with the increasing angle of incidence.

Finally, we use the stochastic techniques described in Chapter 6 to create four pseudo-wells based on the clay content, porosity, and saturation observed in the original data. The goal was to replace the sand in the middle of the interval with shaly sand by increasing the clay content on average by 0.25. These simulations took into account the relation between the clay content and porosity present in the original data (Figure 7.12, left). Hence, in two of the four scenarios, the porosity in the sand was reduced on average by 0.05, while in the other two scenarios we kept the mean porosity in the sand the same as in the original data. Since we did not observe a clear dependence of saturation on the clay content in the original data (Figure 7.12, right), we gradually increased the water saturation in the sand by 0.60. This saturation increase is appropriate for modeling shaly sand with non-commercial amounts of gas. The soft-sand model was then used to compute the density and elastic properties profiles for each of these four scenarios and synthetic seismic gathers were computed (Figures 7.13 to 7.16). The details of the scenarios are given in the captions.

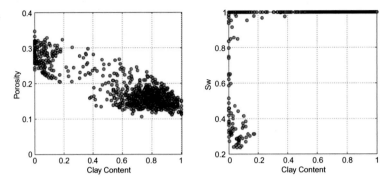

Figure 7.12 Porosity versus clay content (left) and saturation versus clay content (right) in the well under examination.

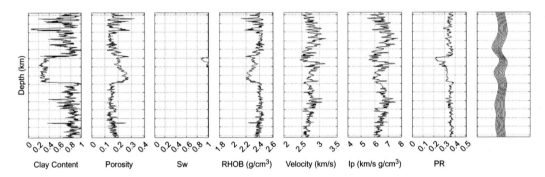

Figure 7.13 A stochastic realization based on the original data from the well examined in this section. The clay content in the sand is increased on average by 0.25 while the porosity is reduced on average by 0.05. The thin shale layers separating the gas from wet sand in the original data are removed. Frequency is 30 Hz.

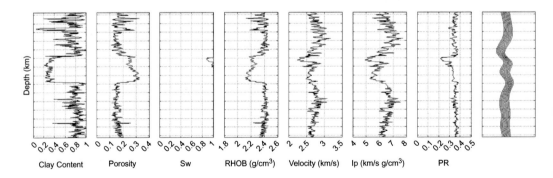

Figure 7.14 Same as in Figure 7.13 but without porosity reduction in the sand.

The results indicate that although the AVO response from shaly sand with small amounts of gas is qualitatively similar to that observed for the original data, it is noticeably weaker. The small reduction in the sand's porosity acts to dim the events at the top of the gas sand as well as at the transition from wet sand to shale. The presence

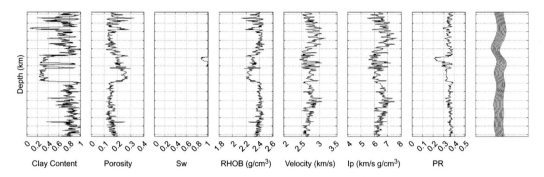

Figure 7.15 Same as in Figure 7.13 but keeping the thin shale layers separating the gas from wet sand in the original data.

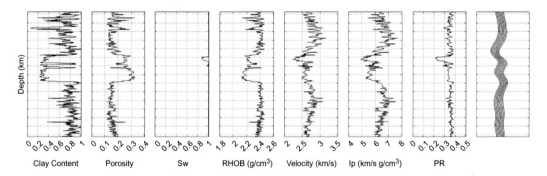

Figure 7.16 Same as in Figure 7.14 but keeping the thin shale layers separating the gas from wet sand in the original data. No porosity reduction in the sand.

of the thin shale layers between the gas and wet sand practically does not affect the amplitude.

Although these stochastic alterations of the original data do not result in dramatic changes in the amplitude of the reflections, the modeling exercises are still important: without conducting such simulations we could not have arrived at this conclusion.

7.2   Consolidated cemented gas sand

Figure 7.17 shows the well log curves for another gas well. A massive gas sand interval is followed by relatively thin gas sand layers separated by shale layers. The difference between these data and the well examined in the previous subsection is that here the sand has smaller porosity and much higher velocity. While in the previous example, the velocity in the gas sand is about 2.5 km/s, here it reaches 4.0 km/s. The difference between the velocities in the shale in the well examined in the previous section and the present well is not that large. In the previous example, the shale velocity was between 2.5 and 3.0 km/s while here it is about 3.0 km/s. This means that the shale in the well under examination is

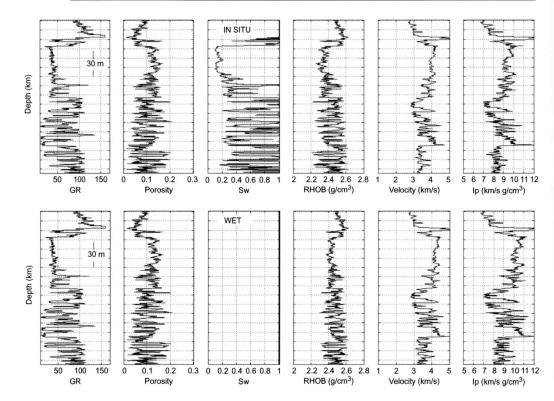

Figure 7.17 Same as Figure 7.1 but for the gas well with consolidated and cemented sand.

slightly more compacted than the shale in the previous example but, in contrast, the sand is not only more compacted but also stiffer due to the onset of diagenetic cementation.

After conducting the V_p-only fluid substitution, we obtain the wet-rock velocity (Figure 7.17, bottom). Because the sand is stiff, fluid substitution from gas to wet does not significantly affect V_p and I_p, unlike in the previous example.

The in situ as well as wet-rock impedance is plotted versus porosity in Figure 7.18 and color-coded by GR. The constant-cement model curves are plotted on top of the wet-rock cross-plot. As explained in Chapter 2, this model has the same functional form as the soft-sand model, except for the artificially high coordination number to account for diagenetic cementation. The parameters used in the respective soft-sand model are: 50 MPa differential pressure; 0.40 critical porosity; 20 coordination number; and 1 shear-modulus correction factor. The bulk modulus of water is 2.30 GPa and its density is 0.96 g/cm³. The bulk modulus and density of gas are 0.08 GPa and 0.23 g/cm³, respectively. This model appears to describe the data with reasonable accuracy (Figure 7.18, right).

As in the previous example, the clay content was computed by running the constant-cement model in reverse. This clay content curve is shown in Figure 7.19. It follows the shape of the GR curve except for the top shale interval, which is too soft to be modeled by the constant-cement model. This is the interval whose P-wave impedance falls below the 100% clay content model curve in Figure 7.18 (right).

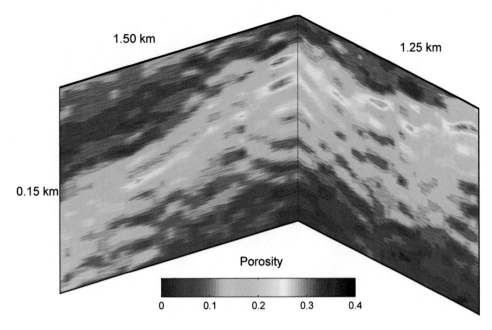

Figure 6.17 Simulated porosity volume (as explained in the text) displayed in two orthogonal cross-sections. The horizontal and vertical distances are marked along the respective directions.

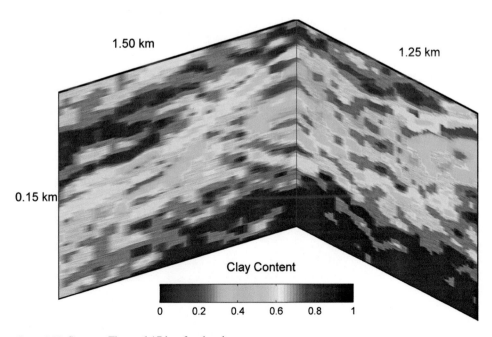

Figure 6.18 Same as Figure 6.17 but for the clay content.

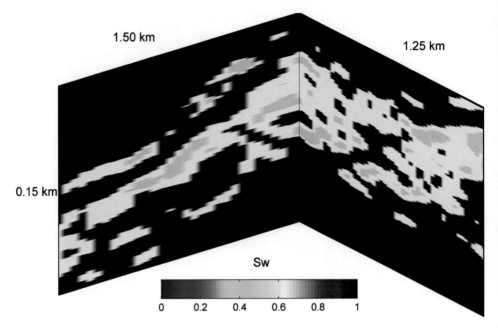

Figure 6.19 Same as Figure 6.17 but for water saturation.

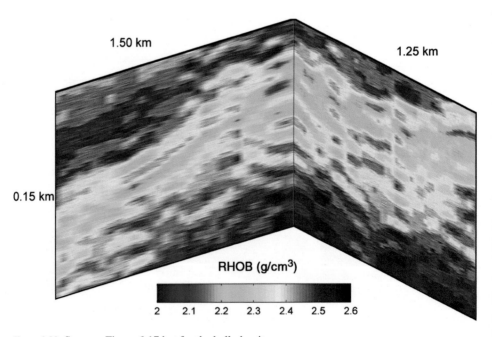

Figure 6.20 Same as Figure 6.17 but for the bulk density.

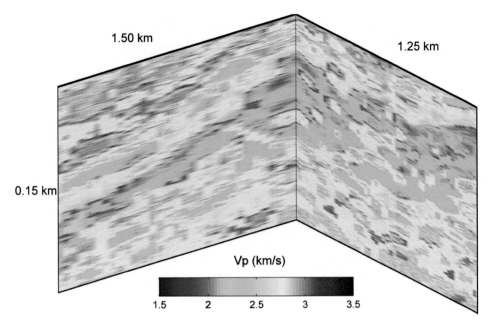

1.50 km

1.25 km

0.15 km

Vp (km/s)

1.5   2   2.5   3   3.5

Figure 6.21 Same as Figure 6.17 but for the *P*-wave velocity.

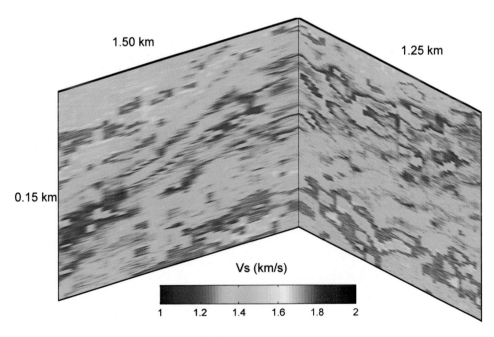

1.50 km

1.25 km

0.15 km

Vs (km/s)

1   1.2   1.4   1.6   1.8   2

Figure 6.22 Same as Figure 6.17 but for the *S*-wave velocity.

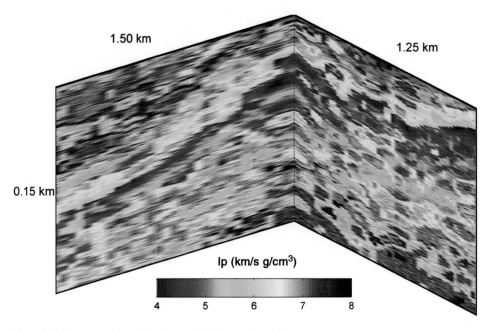

Figure 6.23 Same as Figure 6.17 but for the *P*-wave impedance.

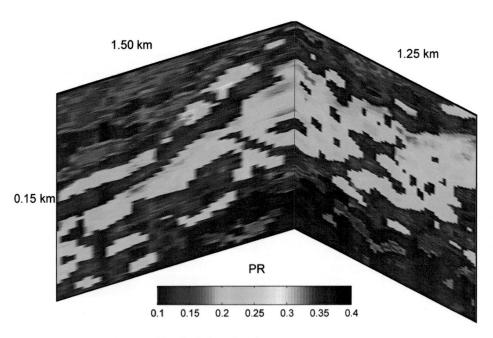

Figure 6.24 Same as Figure 6.17 but for Poisson's ratio.

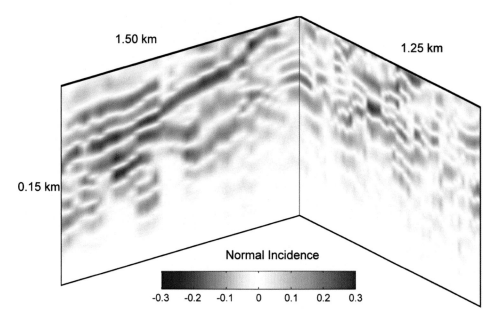

Figure 6.25 Same as Figure 6.17 but for normal incidence amplitude.

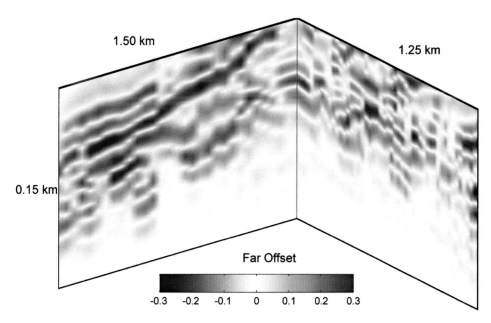

Figure 6.26 Same as Figure 6.17 but for far-offset amplitude.

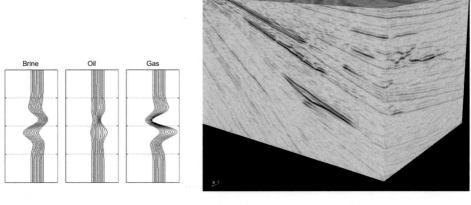

Figure 11.1 A detail of interpretation workflow: synthetic scenario-based seismic gathers to be compared to the actual seismic anomaly in real data cube. In the real-seismic display, red is for negative while blue is for positive amplitude.

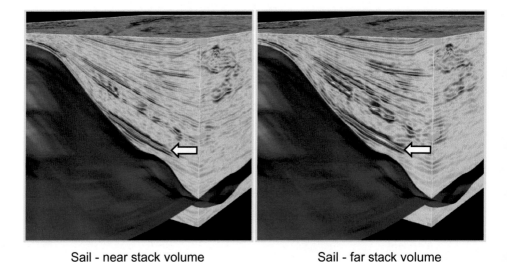

Sail - near stack volume   Sail - far stack volume

Figure 12.1 AVO volumes for near and far stacks. SAIL (seismic approximate impedance log) stack is obtained by integrating the appropriate seismic trace with respect to the travel time (Waters, 1992). Red is for negative while blue is for positive amplitude.

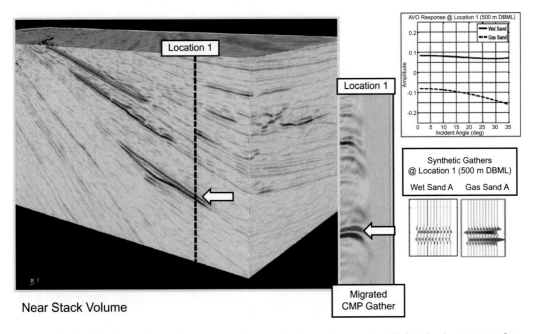

Figure 12.2 Comparison of near-stack volume, migrated gather, and synthetic seismic responses for a wet and gas clean sands scenarios at 500 m DBML (Gutierrez and Dvorkin, 2010).

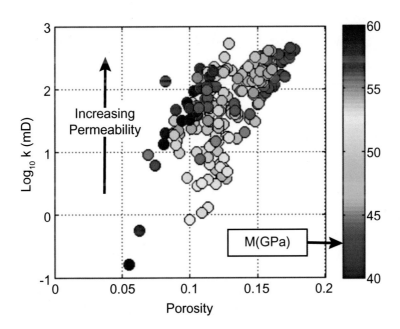

Figure 13.19 Permeability versus porosity color-coded by the compressional modulus. After Grude *et al.* (2013).

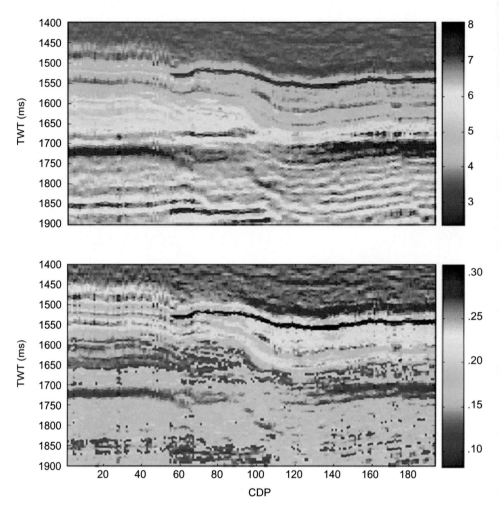

Figure 18.1 Top: Seismically derived impedance section. Bottom: Porosity section obtained by applying Eqs (18.1) and (18.2) to the impedance section. After Dvorkin and Alkhater (2004). The units of the impedance are km/s g/cm³.

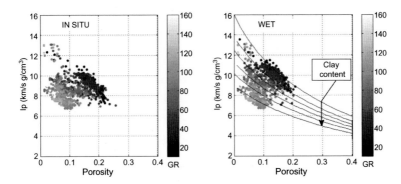

Figure 7.18 The *P*-wave impedance versus porosity for the interval shown in Figure 7.17, color-coded by GR. Left: In situ conditions. Right: Wet-rock impedance computed from the in situ elastic properties by fluid substitution. The model curves on the right are from the constant-cement model as explained in the text.

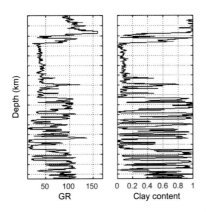

Figure 7.19 GR (left) and clay content curves for the well under examination.

Using this clay content, we apply the constant-cement model to the entire interval at in situ and wet conditions and achieve an accurate match between the measured and modeled density, velocity, and impedance (Figure 7.20).

Now that we are confident that the constant-cement model accurately predicts the elastic properties, we can use it for V_s prediction and synthetic gather generation. As before, we use a 30 Hz Ricker wavelet. The synthetic seismic results shown in Figure 7.21 indicate that the amplitude from the gas-filled interval is very similar to that from the same but wet interval, which is expected in stiff cemented sand. This means that seismic interpretation in the case under examination has to rely on geological and depositional considerations. Rock physics by itself can hardly resolve the difference between the gas and wet intervals.

An AVO modeling applet based on the constant-cement model established by rock physics diagnostics of this interval is shown in Figure 7.22. In this modeling we assumed constant water saturation of 0.20 in the gas sand. The clay content in the sand

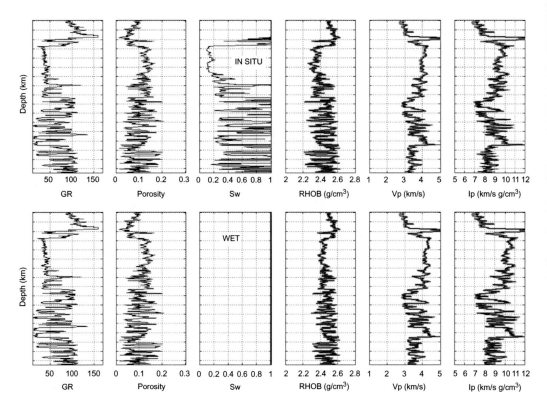

Figure 7.20 Measured and modeled elastic properties in the well as in situ (top) and wet-rock (bottom) conditions. In the density, velocity, and impedance tracks, the bold gray curves are for the measured properties while black curves are for the modeled properties.

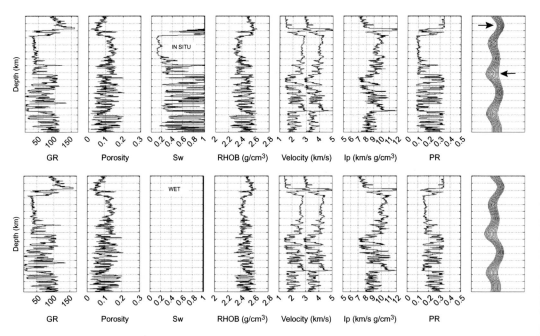

Figure 7.21 Synthetic seismic gather for the in situ (top) and wet (bottom) conditions. The maximum angle of incidence is about 45°. The arrows point to the events at the top and bottom of the massive gas interval. The velocity track shows both the *P*- and *S*-wave velocity.

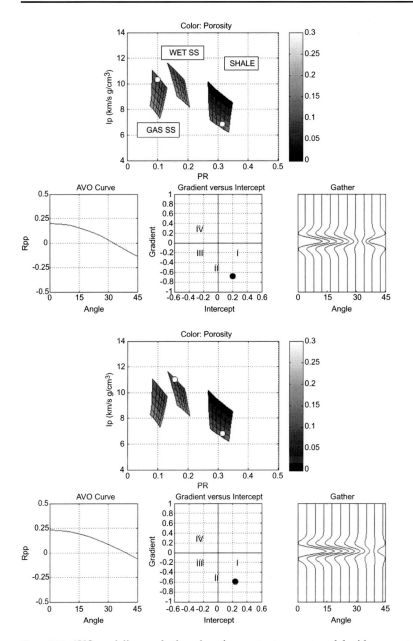

Figure 7.22 AVO modeling applet based on the constant-cement model with parameters described in this section. Left: From relatively soft shale to relatively stiff gas sand. Right: The same but for wet sand.

varies from zero to 0.20, while its porosity varies between 0.10 and 0.20. The clay content in the shale is between 0.70 and 1.00, while its porosity varies between 0.02 and 0.17. In Figure 7.22 we compare reflections at the shale/sand interface for gas and wet sand. The difference between the two is practically indistinguishable.

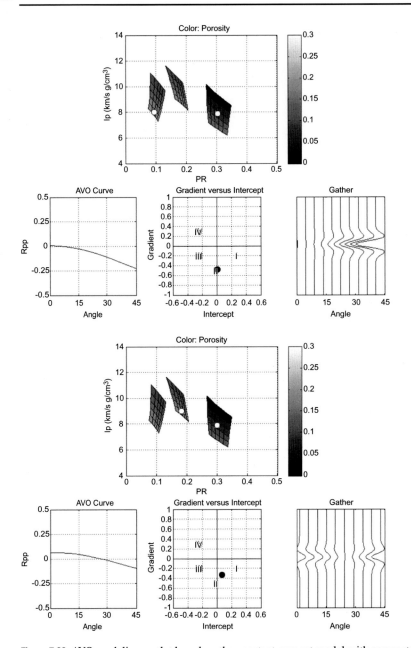

Figure 7.23 AVO modeling applet based on the constant-cement model with parameters described in this section. Left: From relatively soft shale to relatively soft gas sand. Right: The same but for wet sand.

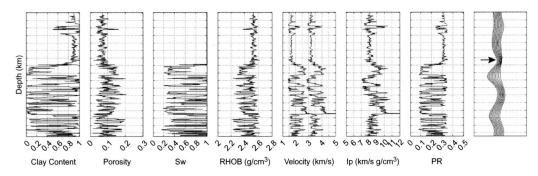

Depth (km)

Clay Content   Porosity   Sw   RHOB (g/cm³)   Velocity (km/s)   Ip (km/s g/cm³)   PR

Figure 7.24 Depth curves and synthetic seismic gather for a hypothetical scenario where the massive gas sand was replaced by shale. The synthetic seismic parameters are the same as in the previous examples of this section. The arrow points at the reflection at the top of the laminated sequence. The velocity track shows both the *P*- and *S*-wave velocity.

In Figure 7.23 we model the reflection between a medium-stiff shale and relatively soft sand. The AVO character changes from the previous example. It is now Class II AVO between the shale and gas sand and very weak Class I between the shale and wet sand.

Let us next explore seismic reflections from the laminated sand/shale sequence in the well under examination by replacing the massive gas sand with shale. To do this, we smoothed the originally established clay content curve by using a running 50-point arithmetic average window, computed the residual as the difference between the original and smoothed clay content curves, and added this residual to a constant clay content of 0.90. The porosity was altered in a similar way. Also, this hypothetical interval above the laminated gas-sand/shale sequence was assumed 100% wet. The same constant-cement model was applied to these hypothetical clay content, porosity, and water saturation inputs. The resulting curves and synthetic seismic gather are shown in Figure 7.24.

The top of the laminated sequence shows a clear Class I AVO response where a normal-incidence positive peak is followed by a number of troughs and peaks due to the reflections inside the laminated sequence.

8 Log shapes at the well scale and seismic reflections in clastic sequences

8.1 Examples of shapes encountered in clastic sequences

The seismic amplitude depends not only on the contrasts of the elastic properties but also on the shape of the elastic property distribution in the subsurface which, in turn, is associated with the underlying rock properties, including the porosity and clay content. In Figure 8.1 we show well data from an offshore oil well that encountered a low-GR hydrocarbon reservoir. The shape on the GR curve is blocky at the bottom but has a gradual low- to high-GR at the top. The porosity and density curves mimic this fining-upwards shape and so does the P-wave impedance. As a result, the impedance contrast at the bottom of the oil-filled interval is larger than at its top and, hence, the amplitude at the bottom is stronger.

It exhibits an AVO response with the normal-incidence positive amplitude becoming increasingly positive with the increasing offset. The reason for this behavior is the positive contrast of the P-wave impedance as well as Poisson's ratio between the sand and underlying shale (Figure 8.1, top). The character of this reflection would be very different if the sand was wet (Figure 8.1, bottom) as the strong impedance and Poisson's ratio contrasts present in situ virtually disappear. This is the case where a bright spot, especially at far angle incidence, points to a hydrocarbon-filled interval.

A similar and even stronger effect can be observed as caused by a fining-upwards interval with residual gas saturation in an offshore well (Figure 8.2). A similar but somewhat weaker effect is caused by two sand intervals (Figure 8.3). The sand in the lower interval is cleaner than that above it, as indicated by the GR curve. Perhaps because of this, the lower sand has residual gas present (in very small quantities) while the upper sand is 100% wet. These two sands are separated by a shale layer. Altogether, this sand/shale combination produces an impedance profile reminiscent of that displayed for a continuous fining-upwards sequence shown in Figure 8.2. Because of this impedance-profile resemblance, the synthetic seismic amplitude has a similar character (Figure 8.3).

In contrast, synthetic seismic reflections from two closely spaced blocky and triangular-shape gas sand intervals show approximately vertically symmetrical amplitude (Figure 8.4).

The synthetic seismic gathers shown in Figures 8.2 and 8.3 are similar in character but the underlying reasons for this similarity are different. This result once again

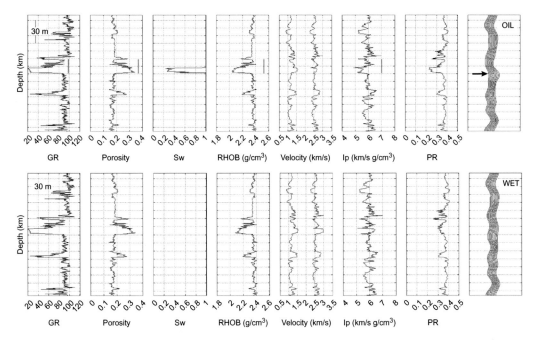

Figure 8.1 Depth curves and synthetic seismic gather for an offshore oil well for the in situ (top) and wet (bottom) conditions. The vertical bars in the in situ display indicate the fining-upwards interval and the arrow points at the seismic anomaly at the bottom of this interval. The synthetic traces were produced using the 30 Hz Ricker wavelet with the maximum incident angle approximately 45°. The velocity track shows both the *P*- and *S*-wave velocity.

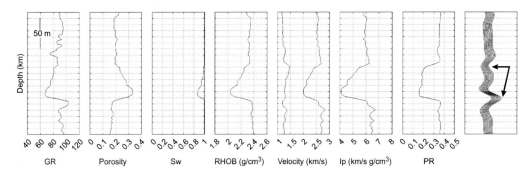

Figure 8.2 Same as Figure 8.1 but for an offshore gas well with residual gas saturation. The synthetic traces were produced using the 30 Hz Ricker wavelet with the maximum incident angle approximately 45°. The well log curves are smoothed. The arrows in the synthetic seismic display point to the amplitude at the top and bottom of the fining-upwards interval. The velocity track shows both the *P*- and *S*-wave velocity.

stresses the non-unique nature of the seismic reflections if they are used to infer stratigraphic configurations of the subsurface: different configurations can produce the same result. This is why it is so important to bring sedimentological and, in general, geological knowledge into the interpretation loop: such information can help constrain

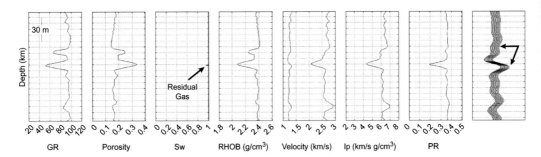

Figure 8.3 Same as Figure 8.2 but for two closely located sand intervals, the lower one with residual gas saturation and the overall impedance profile similar to that in a fining-upwards sequence. The velocity track shows both the *P*- and *S*-wave velocity.

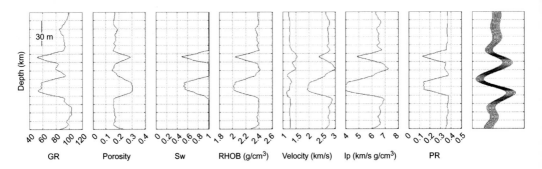

Figure 8.4 Same as Figure 8.2 but for triangular and blocky gas sands.

the modeling variants and, by so doing, mitigate the said non-uniqueness. Still, only geologically meaningful and rock-physics-based forward modeling can open our eyes to the variants possibly responsible for the observed seismic response.

Let us then analyze the rock physics relations between the variables in the fining-upwards section shown in Figure 8.2. Figure 8.5 shows cross-plots between the porosity and impedance and Poisson's ratio and impedance for the in situ as well as wet conditions. The soft-sand model curves superimposed upon the latter two plots indicate that this model is appropriate for describing the interval. The model parameters are: the bulk modulus and density of water are 2.85 GPa and 1.01 g/cm³, respectively; the differential pressure is 30 MPa; the critical porosity is 0.40; the coordination number is 7; and the shear modulus correction factor is 1.

The wet-rock impedance–porosity plot indicates that the data points falling between zero and 20% clay content do not all have the lowest GR values, meaning that in this interval, the clay content does not directly correlate with the GR values. It is rather the total porosity that is most affected by GR. To further analyze the interdependence of the rock properties in this depositional cycle, let us create additional cross-plots, this time including exclusively the wedge-shaped fining-upwards section (Figure 8.6).

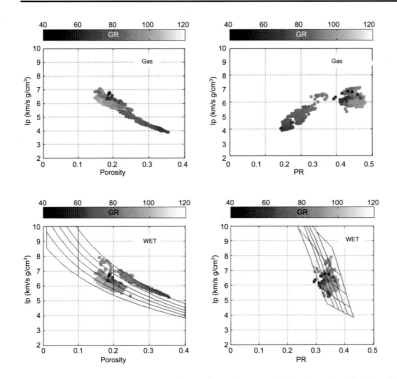

Figure 8.5 Impedance versus porosity (left) and versus Poisson's ratio (right) color-coded by GR. Top: In situ conditions (gas). Bottom: Wet conditions (obtained by fluid substitution). The impedance–porosity and impedance–Poisson's ratio mesh in the bottom plots is from the soft-sand model. The display is similar to the others shown here for rock physics diagnostics purposes.

The cross-plots in Figure 8.6 teach us at least two lessons: (a) the porosity, lithology, and water saturation do not necessarily vary independently of each other (also see Chapter 5) and (b) in spite of a large span of GR in the interval under examination, the clay content varies in a narrow range, between zero and 20%. A diligent student of rock physics may find interesting subtleties in these cross-plots. For example, in the wet-rock impedance versus porosity cross-plot, the data from the highest-porosity lowest-GR interval located at the very bottom of this fining-upwards sequence, fall exactly on the zero-clay soft-sand model curve (see arrow in Figure 8.6, bottom-left), while the rest of the data fall slightly below this model curve. Such subtle variations may tell us more about the history of deposition and, possibly, advance our ability to detect hydrocarbons and infer their amount (saturation). The reader will find an example of such in-depth analysis in Gutierrez (2001) with respect to South America tertiary fluvial clastic deposits.

Let us not worry here about these intriguing subtleties but instead use the lessons learned to construct a pseudo-well with an idealized fining-upwards cycle. Specifically, we assume that the background wet shale has porosity of 0.20 and clay content of 0.80.

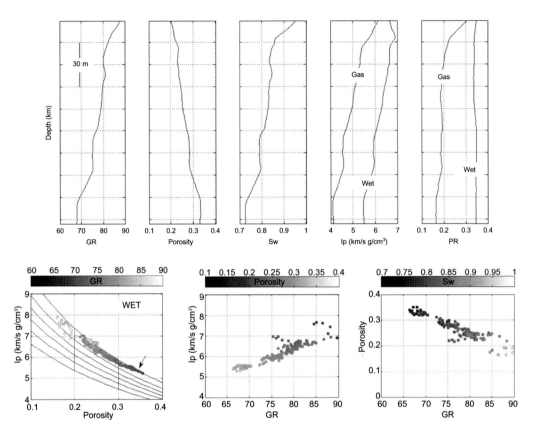

Figure 8.6 Zoom on the fining-upwards section. Top: Well log curves of GR, porosity, water saturation, the *P*-wave impedance (in situ and wet conditions) and Poisson's ratio (in situ and wet conditions). Bottom: Wet-rock impedance versus porosity, color-coded by GR with the soft-sand model mesh superimposed. Wet-rock impedance versus GR, color-coded by porosity. Porosity versus GR color-coded by water saturation.

In the fining-upwards wedge, the clay content is constant at 0.10, while the total porosity gradually decreases from about 0.40 at the bottom to 0.20 at the top. Also, the gas saturation in the sand is constant at 0.80. The bulk moduli of gas and water are 0.041 and 2.65 GPa, respectively, while their respective densities are 0.16 and 1.00 g/cm³.

To compute the elastic properties, we use the soft-sand model with the input parameters the same as used earlier in this section when analyzing real well data. The synthetic seismic gather was generated using a 30 Hz Ricker wavelet with the maximum angle of incidence about 45° (Figure 8.7). Here, as in the real case examined earlier, we observe a strong negative amplitude at the bottom of the wedge-shaped sand and a much smaller amplitude at its top. This simplified pseudo-well produces almost the same seismic response as the real one. Bearing this in mind, let us move to extensive pseudo-well modeling and synthetic seismic generation.

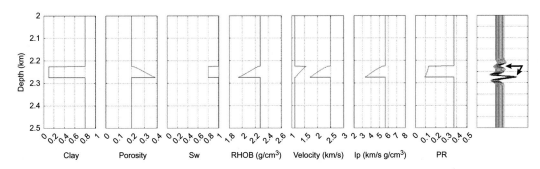

Figure 8.7 Same as Figure 8.2 but for a pseudo-well with a fining-upwards cycle with low gas saturation. The velocity track shows both the *P*- and *S*-wave velocity.

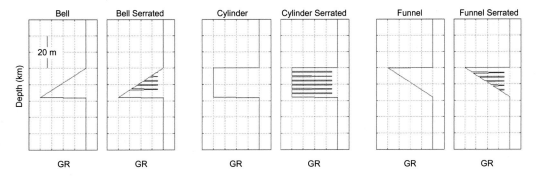

Figure 8.8 Principal GR shape classes in clastic sediment (Rider, 2002). Similar shapes can be observed in the spontaneous potential (SP) data.

8.2 Typical shapes and pseudo-wells in clastic sequences

Rider (2002) describes six principal vertical GR shapes in clastic sediments, the bell (fining-upwards); serrated bell; cylinder (blocky); serrated cylinder; funnel (coarsening-upwards); and serrated funnel (Figure 8.8). Our goal is to compute the elastic properties for such intervals and then generate the respective synthetic gathers.

In the preceding section, we show that although the GR values in the fining-upwards interval almost monotonically increase upwards, the impedance values can only be matched by the soft-sand model with the clay content below 20%. Rider (2002) explains that increasing GR does not necessarily mean increasing clay content. It can be triggered by the presence of other radioactive elements, such as feldspar and mica. To simplify the situation, we assumed that the mineral composition is only quartz and clay with the elastic moduli and density values taken from Table 2.1. We then assumed constant 10% clay content in the fining-upwards sequence in the pseudo-well and produced a synthetic seismic gather similar to that produced with the real well data.

Table 8.1 *Variability of the elastic moduli and densities of minerals (Mavko et al., 2009).*

Mineral	Bulk Modulus (GPa)	Shear Modulus (GPa)	Density (g/cm^3)
Kaolinite	1.5	1.4	1.58
Gulf Clay A	25.0	9.0	2.55
Gulf Clay B	21.0	7.0	2.60
Muscovite A	61.5	41.1	2.79
Muscovite B	42.9	22.2	2.79
Muscovite C	52.0	30.9	2.79
Plagioclase	75.6	25.6	2.63
"Average" Feldspar	37.5	15.0	2.62
Quartz A	37.0	44.0	2.65
Quartz B	36.6	45.0	2.65
Quartz C	37.9	44.3	2.65

This is just one way of dealing with the uncertainty in mineralogy. Generally, it is prudent to first determine from well and core data which is the simplest mineralogy that allows us to explain and quantify well and laboratory data at a prospect under examination and then use this composition for pseudo-wells and synthetic seismic generation. Such simplifications, however, do not guarantee that all possible occurrences in the subsurface will be adequately described.

Rock physics modeling of the elastic properties necessarily requires the mineralogy. In spite of the availability of sophisticated logging tools, such as the spectral gamma-ray tool, a realistic and consistent mineralogical composition is difficult to obtain, especially in the context of elastic-property modeling. Even if the mineral balance is available, we still need to select appropriate elastic moduli and density values for individual minerals. These constants, especially for clay, may strongly depend on the type of mineral. Consider, for example, a table of mineral constants reproduced from Mavko *et al.* (2009) where the clay, mica, feldspar, and even quartz constants span a noticeable range (Table 8.1). The question that faces a modeler is which set of constants to use.

In practice, it is important to select a set of constants that allow us to model the data to the best of our ability and then refrain from varying these constants within a prospect and, especially, versus depth. The upside of rock physics modeling and diagnostics is that any mineralogy and any constants can be used in the effective-medium models. The downside is the enormity of possible variants. Our recommendation is to select mineralogical variants and the elastic and density constants in the way that is "as simple as possible, but not simpler," the latter meaning that such selection used in rock physics and seismic modeling has to explain the available field data and observations.

With this in mind, we proceed with modeling the scenarios shown in Figure 8.8 using the mineral constants from Table 2.1. We will assume that the clay content in the background shale is a constant 80%, while it linearly varies, proportional to the GR, from 5% to 80% in the sand sections. The porosity in the shale will be constant at 0.15, while the porosity at the cleanest sand point is 0.25 and linearly varying in the sand

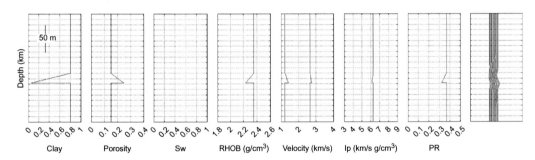

Figure 8.9 The display is the same as used in Figure 8.7. Fining-upwards sequence. The clay content and porosity are computed as explained in the text. 100% wet case. Soft sand model. 30 Hz Ricker wavelet with the maximum angle of incidence about 45°.

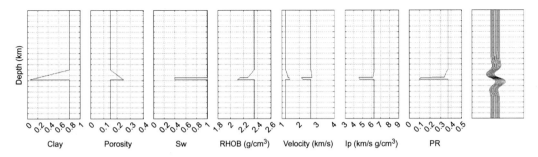

Figure 8.10 Same as Figure 8.9 but for gas sand. Soft-sand model.

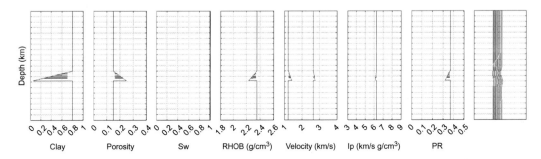

Figure 8.11 Same as Figure 8.9 but for serrated fining-upwards sequence. 100% wet case. Soft-sand model.

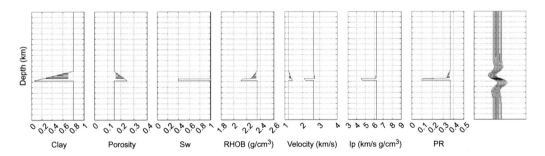

Figure 8.12 Same as Figure 8.11 but for gas sand. Soft-sand model.

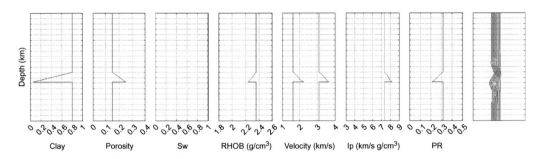

Figure 8.13 Same as Figure 8.9 but using the stiff-sand model.

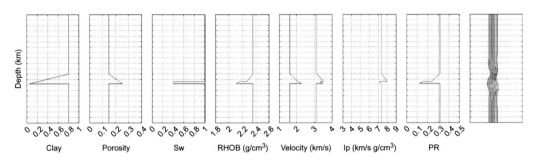

Figure 8.14 Same as Figure 8.13 but for gas sand (the stiff-sand model).

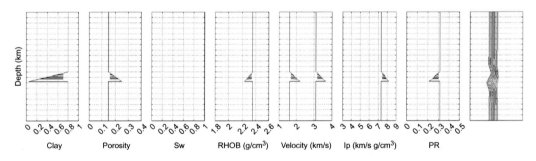

Figure 8.15 Same as Figure 8.13 but for serrated shape (the stiff-sand model).

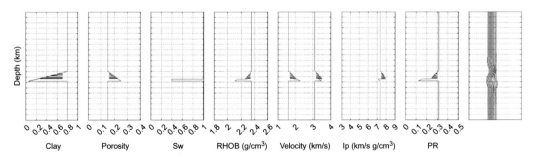

Figure 8.16 Same as Figure 8.15 but for gas sand (the stiff-sand model).

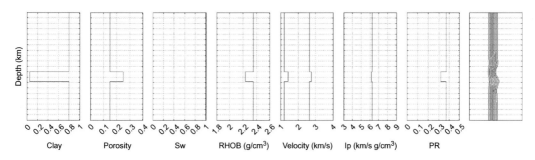

Figure 8.17 Blocky (cylinder) shape. 100% wet case. The soft-sand model.

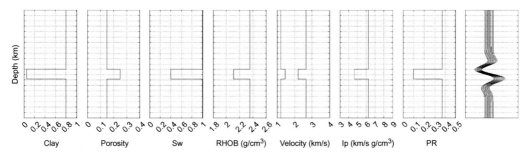

Figure 8.18 Blocky (cylinder) shape. Gas-sand case. The soft-sand model.

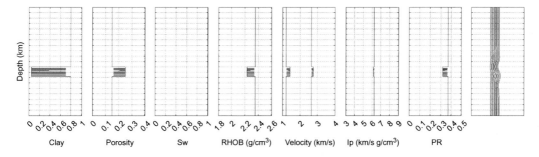

Figure 8.19 Same as Figure 8.17 but for serrated shape. 100% wet case. The soft-sand model.

proportionate to GR between 0.25 and 0.15. We will model the (a) wet-rock case and (b) gas-sand case, the latter with placing a constant water saturation of 40% where the clay content falls below 20%. We will also use two extreme rock physics models: the soft-sand model giving the lowest elastic properties and the stiff-sand model giving the highest elastic properties and apply either of these models throughout the entire interval under examination. The constants used in both models are as follows: 30 MPa for the differential pressure; 0.40 for the critical porosity; 6 for the coordination number; and 1 for the shear-modulus correction factor. The pore-fluid properties will be the same as used in the preceding section: the bulk moduli of 0.041 and 2.65 GPa for gas and water, respectively with the accompanying densities of 0.16 and 1.00 g/cm³.

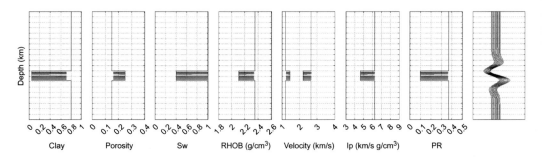

Figure 8.20 Same as Figure 8.18 but for serrated shape. Gas-sand case. The soft-sand model.

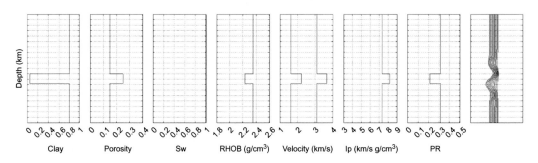

Figure 8.21 Same as Figure 8.17 but using the stiff-sand model. 100% wet case.

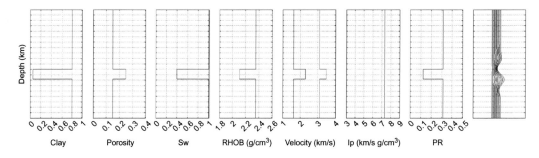

Figure 8.22 Same as Figure 8.21 but for gas-sand case (the stiff-sand model).

The depth curves and synthetic gathers thus computed for the fining-upwards shape are shown in Figures 8.9 to 8.16. For the frequency and thickness used in these examples, the serrated shape produces synthetic gathers practically the same as those generated for the smooth shape. However, the response varies between the 100% wet and gas cases, and also strongly depends on the rock physics model used (the soft versus stiff case).

The results for the blocky sand (cylinder) are shown in Figures 8.17 to 8.24, while those for the coarsening-upwards sequence (funnel) are shown in Figures 8.25 to 8.32. Evident is the influence of the shape and presence of gas, as well as the sediment type (soft versus stiff) on the synthetic seismic response. At the same time, the presence of

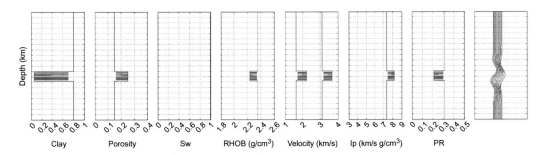

Figure 8.23 Same as Figure 8.21 but for serrated profile (100% wet case, the stiff-sand model).

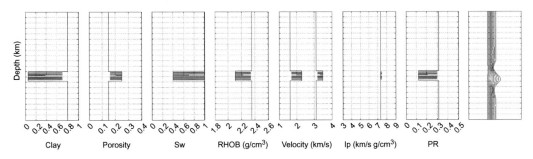

Figure 8.24 Same as Figure 8.22 but for serrated profile (gas-sand case, the stiff-sand model).

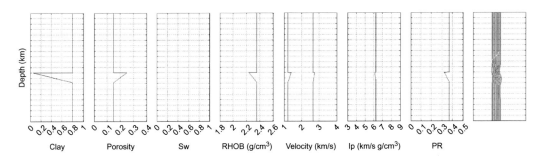

Figure 8.25 Coarsening-upwards sequence. 100% wet case. The soft-sand model.

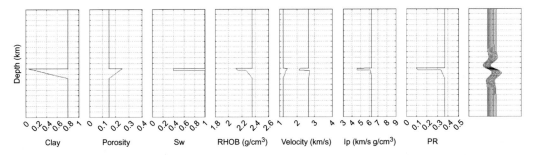

Figure 8.26 Same as Figure 8.25 but for gas-sand case. The soft-sand model.

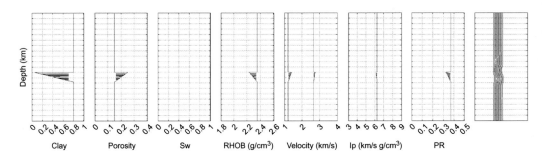

Figure 8.27 Coarsening-upwards serrated sequence. 100% wet case. The soft-sand model.

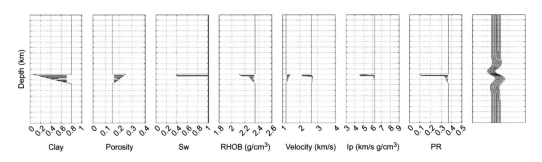

Figure 8.28 Coarsening-upwards serrated sequence. Gas-sand case. The soft-sand model.

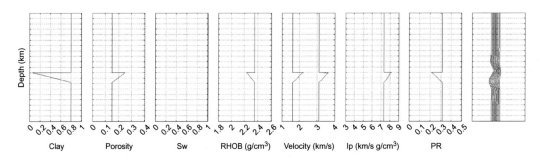

Figure 8.29 Coarsening-upwards sequence. 100% wet case. The stiff-sand model.

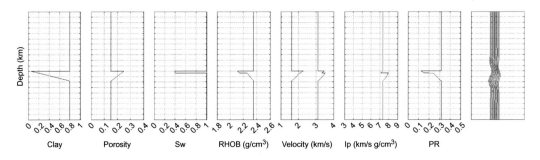

Figure 8.30 Coarsening-upwards sequence. Gas-sand case. The stiff-sand model.

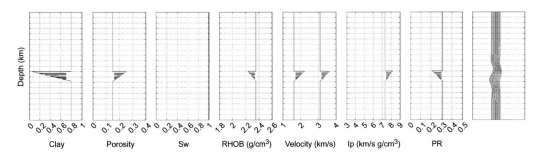

Figure 8.31 Coarsening-upwards serrated sequence. 100% wet case. The stiff-sand model.

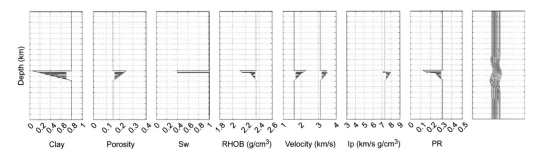

Figure 8.32 Coarsening-upwards serrated sequence. Gas-sand case. The stiff-sand model.

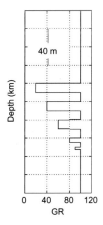

Figure 8.33 GR shape for delta border progradation.

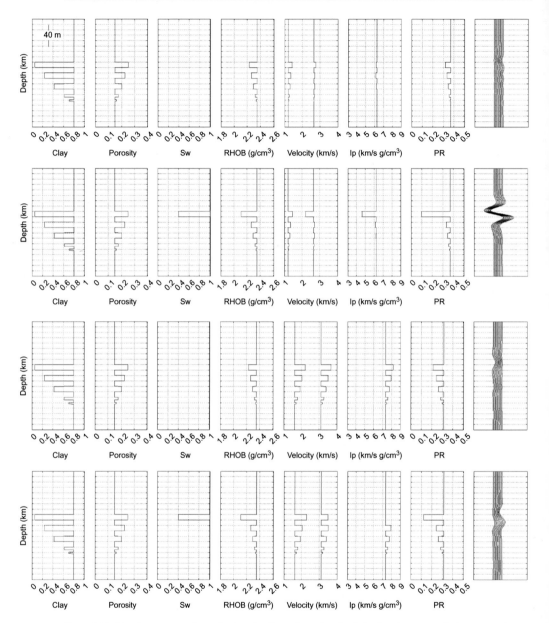

Figure 8.34 Depth curves and synthetic seismic gathers for the delta border progradation scenario. From top to bottom: wet soft sand; gas soft sand; wet stiff sand; and gas stiff sand.

serrations hardly affects the response in the fining-upwards and coarsening-upwards cases. However, the serrations affect the response of the blocky sand (compare, e.g., Figure 8.21 and 8.23). This effect is expected, as the shale serrations in the blocky sand soften the overall impedance and, at the same time, slightly increase Poisson's ratio at the seismic scale.

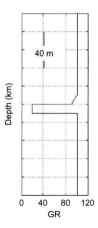

Figure 8.35 GR shape for transgressive marine shelf.

According to Rider (2002), the GR shapes correspond to certain depositional environments. For example, in the deltaic–fluvial environment, a fining-upwards shape indicates a channel point bar. In the same environment, delta border progradation may result in a set of relatively thick sand layers with GR gradually increasing from top to bottom (Figure 8.33).

Using the same approach for building the clay content and porosity profiles as in the previous examples, we can create the elastic properties and synthetic seismic response for this case for wet and gas sand, using the soft-sand (uncemented and poorly sorted) and stiff-sand (well sorted cemented) models (Figure 8.34). The seismic response in this case strongly depends on the stiffness of the sediment as well as on the presence of gas.

Once again, according to Rider (2002), a transgressive marine shelf environment may feature a nearly-blocky sand graduating into the shale above with gently increasing GR shape (Figure 8.35). The respective depth curves and synthetic seismic gathers are shown in Figure 8.36. We observe that the subtleties of the well-scale mineralogy and porosity variations are muted at the seismic scale. This does not mean that the rock-physics-based forward modeling is futile in the context of revealing the subsurface properties from seismic data. Just the opposite, it is crucial for understanding what we can see and discriminate and what we cannot.

A GR shape for a prograding marine shelf environment is shown in Figure 8.37. It is driven by a blocky sand gradually changing with depth into a series of thinner higher-GR intervals. The respective seismic response variants are shown in Figure 8.38.

The deep-sea environment GR shapes translated into the clay content in the same way as in the previous examples in this section are shown in Figures 8.39 to 8.43. In this case, to illustrate the respective seismic responses, we will only use the gas-sand cases and employ a constant-cement model where the only difference from the soft-sand model is in selecting a higher coordination number, 15 in these examples. The specific cases examined here include (a) slope channel (Figure 8.39); (b) inner fan channel (Figure 8.40); (c)

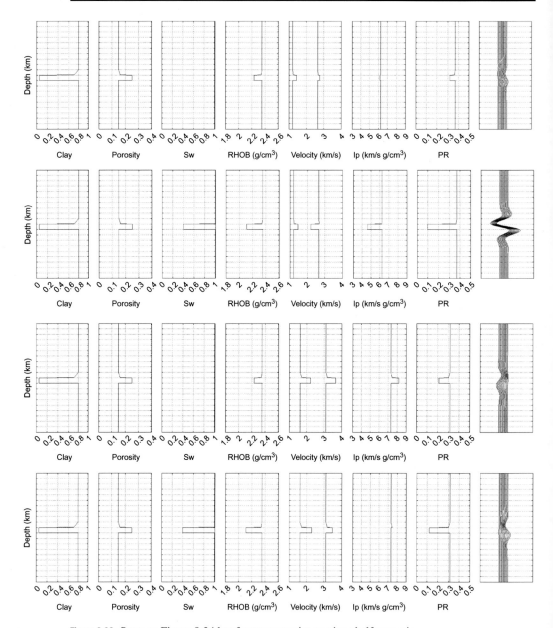

Figure 8.36 Same as Figure 8.34 but for transgressive marine shelf scenario.

middle fan channel (Figure 8.41); (d) supra-fan depositional lobes (Figure 8.42); and (e) distal basin plain (Figure 8.43); all shapes drawn following Rider (2002).

In these examples we used fixed clean-sand and pure-shale mineralogy and porosity. In practical applications, these quantities, as well as the mineralogy have to be tied to concrete geological circumstances, including the overburden and differential pressure and temperature.

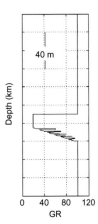

Figure 8.37 GR shape for prograding marine shelf.

The final example in this section pertains to the effect of the sand mineralogy on the elastic properties and seismic response. Specifically, we will examine a blocky sand imbedded into a shale background. The mineralogy of the shale is 80% clay and 20% quartz. Its porosity is constant at 0.15. For the sand we examine two variants: (a) 5% clay and 95% quartz and (b) 5% clay and 95% "average" feldspar with elastic moduli and density from Table 8.1. Replacing quartz with feldspar in this exercise reflects the variation between pure quartz and arkosic sand. For the quartz and clay we use the usual constants: 36.6 and 45.0 GPa for the bulk and shear moduli of quartz, respectively, and 2.65 g/cm³ for its density; and 21.0 and 7.0 GPa for the bulk and shear moduli of clay, respectively, and 2.58 g/cm³ for its density. The sand has constant porosity of 0.25. The elastic properties in the interval are computed using the constant-cement model (the soft-sand model with coordination number 15). All other parameters are the same as used in the previous examples. The results are shown in Figures 8.44 and 8.45.

This mineralogy substitution results in a stronger amplitude for feldspar sand compared to the quartz sand for both the wet- and gas-sand cases. Also, the character of the gather varies between the two mineralogies: in the quartz/wet case, we observe a weak AVO Class I reflection with the normal-incidence reflection positive, gradually decreasing with the increasing angle of incidence and becoming slightly negative at about 45°. In contrast, wet feldspar produces a weak AVO Class IV response with the normal-incidence trough becoming smaller with the increasing angle of incidence (Figure 8.44).

The gas-sand reflections are stronger in the feldspar case and have a different character to that for the quartz sand: in the former case we observe a clear AVO Class III with the reflection becoming increasingly negative with the increasing angle of incidence. Although the feldspar–sand amplitude is stronger than that for the quartz–sand case, the AVO effect is much less pronounced in the former than in the latter.

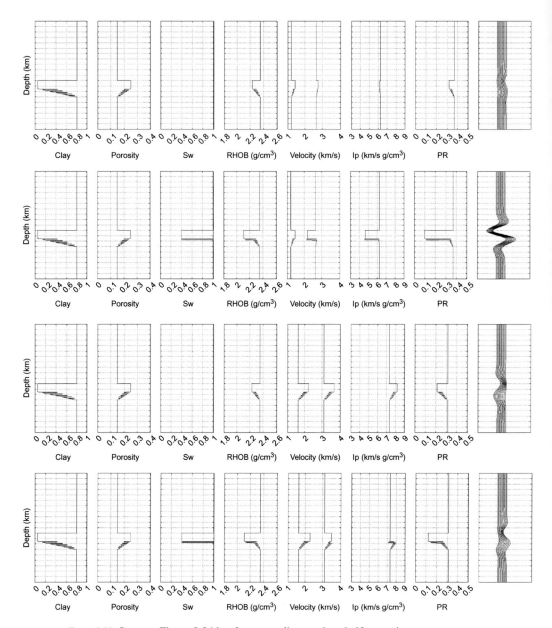

Figure 8.38 Same as Figure 8.34 but for prograding marine shelf scenario.

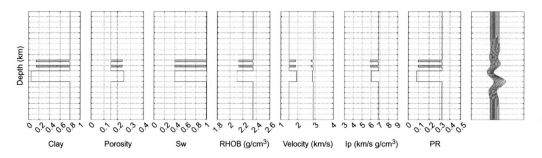

Figure 8.39 Deep sea environment. Slope channel with gas sand and using the constant-cement model (the soft-sand model with coordination number 15).

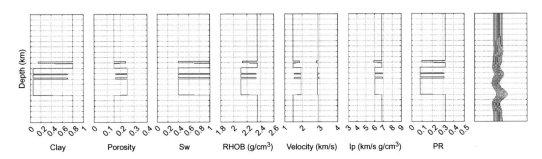

Figure 8.40 Same as Figure 8.39 but for inner fan channel.

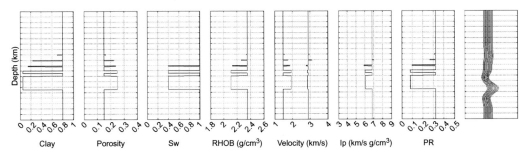

Figure 8.41 Same as Figure 8.39 but for middle fan channel.

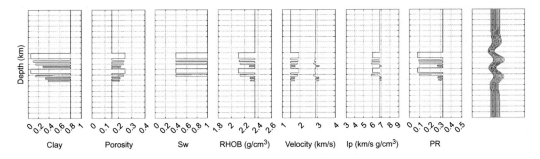

Figure 8.42 Same as Figure 8.39 but for supra-fan depositional lobes.

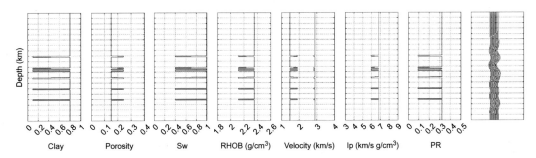

Figure 8.43 Same as Figure 8.39 but for distal basin plain.

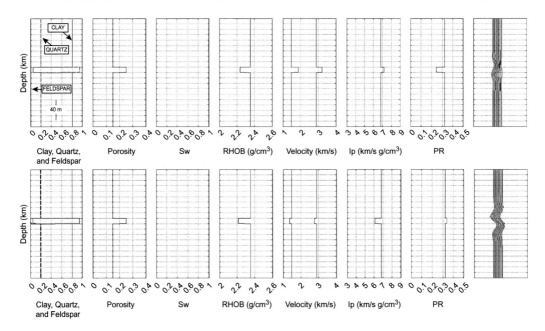

Figure 8.44 Top: Blocky sand with quartz and no feldspar. Wet conditions. Constant-cement model. Bottom: Same but for quartz replaced by feldspar in the sand. In the first track, the thin black curve is for clay content; the bold gray curve is for quartz content; and the dashed black curve is for feldspar content.

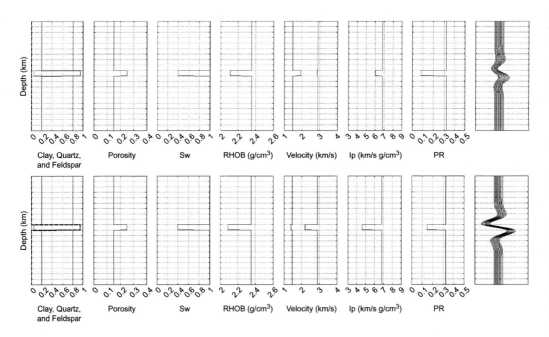

Figure 8.45 Same as Figure 8.44 but for gas sand.

9 Synthetic modeling in carbonates

9.1 Background and models

Hydrocarbons from carbonate reservoirs constitute a significant (if not dominant) part of oil and gas resources worldwide. Yet, compared to sandstones, the rock physics of carbonates is much less developed. The main reason is, arguably, the fact that the pore-space geometry in carbonates is much more complex than in siliceous clastic rock. Wang (1997) lists a number of pore-space types encountered in carbonate rocks and relates them to the elastic-wave velocity in such samples. Some of the pore types are:

Intercrystalline porosity, by definition, is the pore space between crystals of approximately the same size, while **interparticle** porosity is that between particles of any size. Such pores are irregularly shaped and angular, and, therefore, are easy to deform. Hence, both V_p and V_s exhibit strong sensitivity to the differential pressure. **Moldic** pores are created by selective dissolution of individual constituents and have regular shapes. **Intraparticle** pores are located inside individual particles. Such rocks are stiff and their elastic properties weakly depend on the differential pressure. **Vugs** and **channels** are relatively large inclusions resulting from the dissolution of reactive minerals. Rocks with vugs are stiff. Their elastic moduli only weakly depend on the applied stress. Because channel pores are elongated, they make the rock softer and its elastic properties strongly stress-dependent.

A number of empirical velocity–porosity and velocity–density equations have been proposed for carbonates. First and foremost, the classical empirical equations by Wyllie *et al.* (1956) and Raymer *et al.* (1980) can be applied to carbonate rock. Specifically, Wyllie's time average for V_p states that

$$\frac{1}{V_p} = \frac{1-\phi}{V_{ps}} + \frac{\phi}{V_{pf}},$$

(9.1)

where, as before, ϕ is the porosity, V_{ps} is the P-wave velocity in the mineral phase, and V_{pf} is the P-wave velocity in the pore fluid. Clearly, Eq. (9.1) has to be used with caution since in dry rock $V_{pf} \approx 0$ and, hence, the resulting V_p is also zero. It is important to remember that this equation was designed for wet rock samples and should not be used otherwise. Referring to Table 2.1 for the bulk and shear moduli and density of calcite and dolomite, we find that in pure calcite, $V_p = 6.640$ km/s and $V_s = 3.436$ km/s. Likewise, in dolomite, $V_p = 7.347$ km/s and $V_s = 3.960$ km/s.

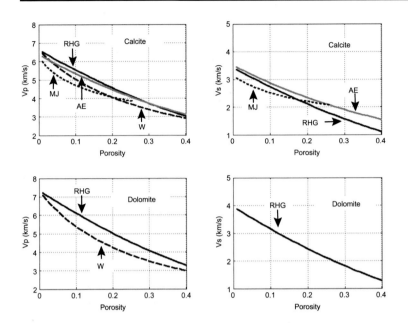

Figure 9.1 Velocity versus porosity for calcite (top) and dolomite (bottom) rock according to Eqs (9.1) to (9.5). The curves are labeled according to the equations used: W for Eq. (9.1); RHG for Eqs (9.2) and (9.3); AE for Eqs (9.4); and MJ for Eqs (9.5). All curves are for wet rock.

The Raymer *et al.* (1980) equation (Chapter 2) is

$$V_p = (1-\phi)^2 V_{ps} + \phi V_{pf}, \tag{9.2}$$

with V_s given by Dvorkin (2008a) as

$$V_s = (1-\phi)^2 V_{ss} \sqrt{\frac{(1-\phi)\rho_s}{(1-\phi)\rho_s + \phi\rho_f}}, \tag{9.3}$$

where V_{ss} is the S-wave velocity in the mineral phase whose density is ρ_s and ρ_f is the density of the pore fluid.

Figure 9.1 shows velocity–porosity cross-plots for pure calcite and pure dolomite wet rock according to Eqs (9.1)–(9.3). The bulk modulus of water in this example is 2.50 GPa and its density is 1.00 g/cm^3.

Anselmetti and Eberli (1997) provide the following velocity–porosity equations based on laboratory measurements performed on carbonate samples from the Bahamas and the Maiella Platform:

$$V_p = 6.393e^{1.80\phi}, \quad V_s = 3.527e^{2.06\phi}, \tag{9.4}$$

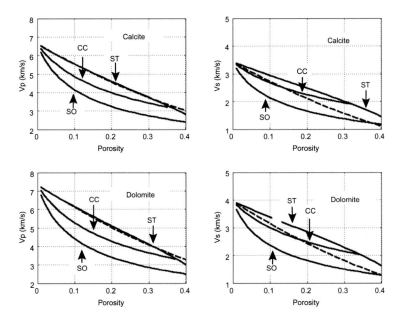

Figure 9.2 Same as Figure 9.1 but with the stiff- (ST), soft- (SO), and constant-cement (CC) curves shown. The RHG curve from Figure 9.1 is shown as a dashed curve. All curves are for wet rock.

where velocity is in km/s. The corresponding curves are also shown in Figure 9.1. The V_p curve falls close to the Eq. (9.2) curve, while the V_s curve deviates from the Eq. (9.3) curve at high porosity.

Marion and Jizba (1997) propose equations relating the bulk (K_{Dry}) and shear (G_{Dry}) elastic moduli of the dry rock frame to its porosity:

$$K_{Dry} = (0.0198 + 0.198\phi)^{-1}, \quad G_{Dry} = (0.0374 + 0.250\phi)^{-1}, \tag{9.5}$$

with the respective curves for wet rock (applying fluid substitution to the bulk modulus given by Eq. (9.5)) plotted in Figure 9.1.

We also have theoretical velocity–porosity models, the soft-sand, stiff-sand, and constant-cement (Chapter 2). These model curves are shown in Figure 9.2. The critical porosity for all three models was 0.45. The coordination number was 6 for the soft- and stiff-sand models and 20 (in the soft-sand model) for the constant-cement model. The differential pressure was 40 MPa and the shear modulus correction factor was 1. The fluid was water with the bulk modulus of 2.50 GPa and density of 1.00 g/cm³.

Ruiz (2009) introduced an effective-medium model for carbonate with porous grains where the effective elastic properties of the grain material were computed using the stiff-sand model or an inclusion model. The elastic moduli thus computed were then assumed to be those of the porous grain material and could be used with any rock physics model. In the same work, it was shown that carbonate data can be approximately

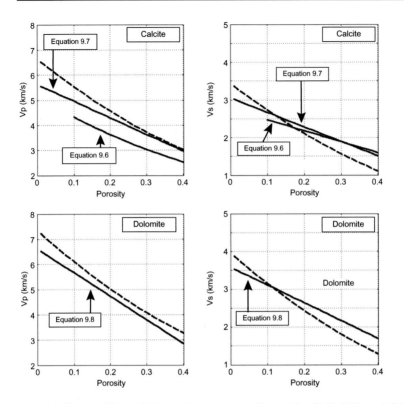

Figure 9.3 Same as Figure 9.1 but with curves according to Eqs (9.6), (9.7), and (9.8), as marked in the plots. The RHG curve from Figure 9.1 is shown as a dashed curve. All curves are for wet rock.

modeled using the differential effective-medium model (DEM) where the inclusions have an aspect ratio of 0.13.

Finally, let us quote statistical equations based on laboratory and well data (Mavko *et al.*, 2009). For chalk samples, water- and oil-saturated,

$$V_p = 5.128 - 8.505\phi + 5.050\phi^2; \quad V_s = 2.766 - 2.933\phi. \tag{9.6}$$

For water-saturated limestone,

$$V_p = 5.624 - 6.650\phi; \quad V_s = 3.053 - 3.866\phi. \tag{9.7}$$

For water-saturated dolomite,

$$V_p = 6.606 - 9.380\phi; \quad V_s = 3.581 - 4.719\phi. \tag{9.8}$$

These three relations are plotted in Figure 9.3 where the Raymer–Dvorkin curves are left for reference.

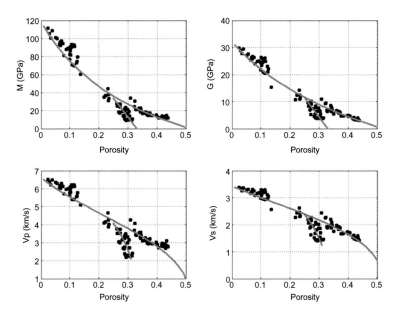

Figure 9.4 Scotellaro *et al.* (2008) carbonate data with model lines (described in the text).

9.2 Laboratory and well data

Scotellaro *et al.* (2008) and Vanorio *et al.* (2011) analyze velocity–porosity trends from laboratory measurements conducted on a large number of carbonate samples extracted from outcrops in Italy. One interesting feature of these data is that the velocities in most of the samples weakly depend on the differential pressure. The reason may be that these samples have existed at atmospheric conditions for so long that the microscopic cracks largely responsible for velocity-stress variation have healed. These dry-rock data are shown in Figure 9.4. The data contain at least two branches: the upper branch smoothly varies between about 0.45 and zero porosity, while the other branch is for softer samples and varies between about 0.30 and 0.25 porosity, catching up with the first branch at 0.25 porosity. According to Vanorio and Mavko (2011), the soft (lower) branch in the data located approximately in the 0.25 to 0.30 porosity range corresponds to microcrystalline (micritic) texture dissolution and formation of large compliant pores or removal of micrite matrix that acts to stiffen the large grain framework.

Figure 9.4 also shows model curves for both branches. The upper-branch curve is the stiff-sand model for dry, pure-calcite rock, with a differential pressure of 1 MPa (most of the measurements were conducted at ambient conditions); a critical porosity of 0.50; coordination number 6; and the shear modulus correction factor 1. To match the lower branch in the data, we simply linearly interpolate the apparent elastic moduli between the lowest and highest porosity in this group of the samples. Then these linear curves in the modulus–porosity domain are translated into velocity–porosity curves

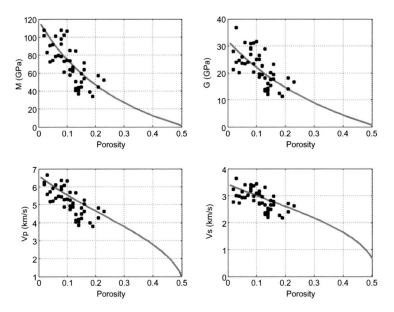

Figure 9.5 Kenter *et al.* (1997) carbonate data with the model curves from the stiff-sand model, the same as in Figure 9.4.

with the bulk density computed for pure calcite with a mineral density of 2.71 g/cm³. The respective equations for the compressional (M_{Dry}) and shear (G_{Dry}) moduli are

$$M_{Dry} = 139 - 422\phi; \quad G_{Dry} = 47 - 143\phi, \tag{9.9}$$

where the moduli are in GPa.

Another relevant dataset is due to Kenter *et al.* (1997). The samples in this study contained both calcite and dolomite, as well as smaller amounts of mica, kaolinite, feldspar, and quartz. The wet-rock measurements were conducted in the laboratory at ultrasonic frequency. Figure 9.5 shows these data at 30 MPa confining pressure. On average, the data can be described by the stiff-sand model curve shown in Figure 9.4. However, there is significant scatter of the data points around this curve, possibly triggered by variations in the pore-space geometry and mineralogy.

Well data from an Ekofisk well dominated by chalk is shown in Figure 9.6 (after Walls *et al.*, 1998). Fluid substitution for 100% wet conditions acts to increase the V_p in the main reservoir by about 18% and increase the Poisson's ratio from approximately 0.25 to 0.35.

The wet-rock compressional and shear moduli are plotted versus porosity in Figure 9.7. The data are matched by the stiff-sand curve computed for wet calcite with a differential pressure of 40 MPa, a critical porosity of 0.55, coordination number 6, and the shear modulus correction factor 2. The bulk modulus and density of water were 2.627 GPa and 0.997 g/cm³, respectively. To adjust the model curve to the data, we had to change the bulk and shear moduli of pure calcite and use 55 GPa and 22 GPa, respectively.

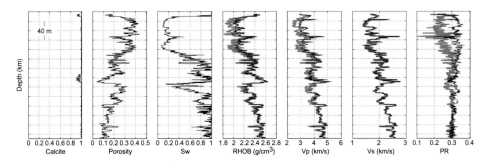

Figure 9.6 Ekofisk well (Walls *et al.*, 1998). In the density, velocity, and Poisson's ratio tracks, bold gray curves are for in situ measurements while black curves are computed curves for 100% wet rock.

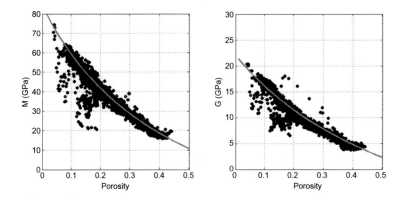

Figure 9.7 Ekofisk well (Walls *et al.*, 1998). Compressional (left) and shear (right) modulus versus porosity for wet rock. The curves are from the stiff-sand model with the mineral properties adjusted as described in the text.

This seemingly ad hoc adjustment indicates that although the functional form used for modeling is appropriate, it has to be anchored by different endpoints. In practical applications it would be wrong to make such adjustments separately for each well. Rather the entire dataset for the field under examination has to be considered and common adjustment constants introduced.

9.3 Pseudo-wells and reflections

Various seismic signatures of carbonates are discussed in Palaz and Marfurt (1997). Sedimentary sequences encountered in and around carbonate reservoirs are discussed and analyzed in Lucia (2007). Here we first present a number of scenarios where a carbonate reservoir is bounded by compacted shale. The shale background has constant porosity 0.15 and clay content 60% (the rest is quartz). The carbonate reservoir can be calcite or dolomite with varying porosity and gas or oil saturation.

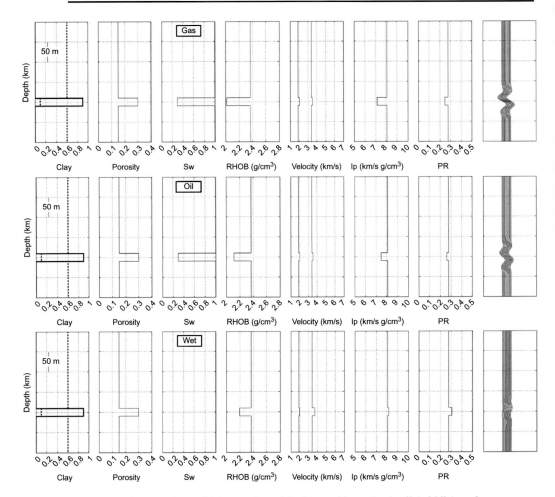

Figure 9.8 Reflections from a high-porosity calcite layer with gas (top), oil (middle), and wet (bottom). In the first track, the clay content is shown by a dotted line while the solid black line is for calcite content. The maximum angle of incidence is about 45°. Frequency is 40 Hz.

To compute the elastic properties in the shale background, we use the stiff-sand model with the differential pressure of 30 MPa, critical porosity of 0.40, coordination number 6, and shear modulus correction factor 1. The same parameters are used in the carbonate layer, except that the critical porosity is now set at 0.50.

Figure 9.8 compares reflections from a 30% porosity calcite layer filled with gas or oil (in both cases $S_w = 0.30$), or fully wet. The fluid properties are from Table 2.2.

In this high-porosity carbonate reservoir we can clearly see the dependence of the amplitude on the pore fluid: the negative amplitude of calcite with gas or oil becomes almost zero where the calcite is filled with water.

Figure 9.9 illustrates the same exercise but with dolomite instead of calcite. The change in mineralogy makes a big difference: there is almost no reflection from dolomite with gas, while we obtain a Class I AVO response from dolomite with oil and

Table 9.1 *Lithofacies, their vertical (top-to-bottom) order, thickness, porosity, and elastic properties for the tertiary limestone buildup scenario. The lithofacies are numbered for display in Figure 9.12.*

Lithofacies	Thickness (m)	Porosity	V_p (km/s)	V_s (km/s)	Density (g/cm³)
Mudrock – 1	50	0.152	2.800	1.288	2.450
Packstone – 2	40	0.080	4.700	2.532	2.550
Wakestone – 3	80	0.250	3.500	1.854	2.270
Grainstone – 4	40	0.100	4.000	2.155	2.520
Quartz-rich packstone – 5	20	0.050	5.400	2.854	2.600
Sandstone – 6	50	0.260	3.200	1.717	2.240

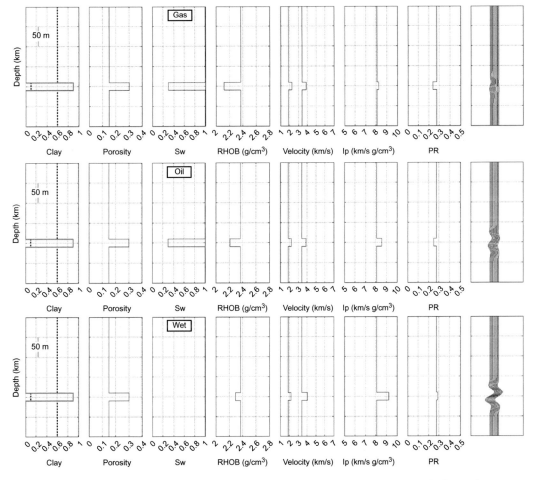

Figure 9.9 Reflections from a high-porosity dolomite layer with gas (top), oil (middle), and wet (bottom). In the first track, the clay content is shown by a dotted line while the solid gray line is for dolomite content. The maximum angle of incidence is about 45°. Frequency is 40 Hz.

Table 9.2 *Lithofacies, their vertical (top-to-bottom) order, thickness, porosity, and elastic properties for the Permian dolostone scenario. The lithofacies are numbered for display in Figure 9.13.*

Lithofacies	Thickness (m)	Porosity	V_p (km/s)	V_s (km/s)	Density (g/cm^3)
Dolowakestone – 1	30	0.050	6.100	3.840	2.780
Dolograinstone – 2	5	0.200	4.800	2.720	2.500
Dolopackstone – 3	30	0.100	5.500	3.130	2.680
Dolowakestone – 1	30	0.050	6.100	3.840	2.780
Dolograinstone – 2	5	0.200	4.800	2.720	2.500
Dolopackstone – 3	30	0.100	5.500	3.130	2.680
Dolowakestone – 1	30	0.050	6.100	3.840	2.780

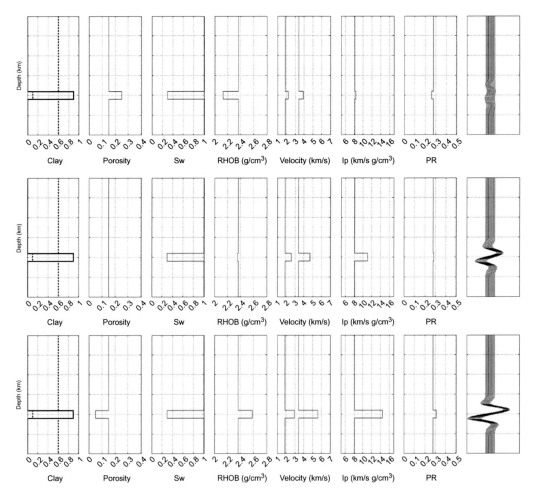

Figure 9.10 Reflections from a gas-filled calcite layer with porosity decreasing from 0.25 (top) to 0.15 (middle) and 0.05 (bottom). Display is the same as in Figure 9.8.

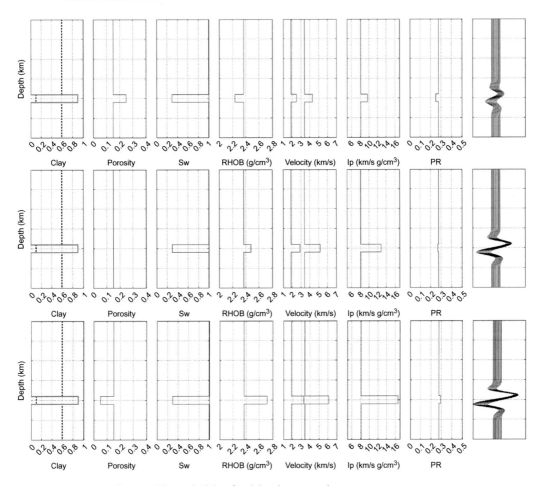

Figure 9.11 Same as Figure 9.10 but for dolomite reservoir.

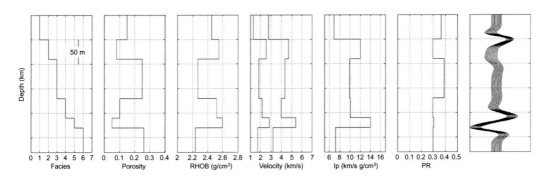

Figure 9.12 Depth curves and synthetic seismic gather for the limestone buildup sequence listed in Table 9.1. Frequency is 40 Hz with maximum incidence angle about 40°.

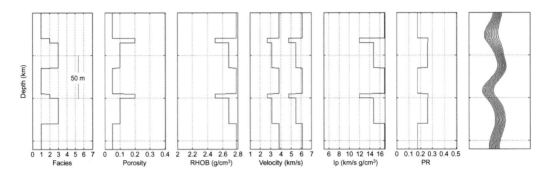

Figure 9.13 Depth curves and synthetic seismic gather for the Permian dolostone sequence listed in Table 9.2. Frequency is 40 Hz with maximum incidence angle about 15°.

water. This class is characterized by a positive reflection for normal incidence with the amplitude gradually reducing with the increasing angle of incidence.

In the next exercise, we will assume that the carbonate reservoir is filled with gas and explore the effect of changing porosity. Specifically, for both calcite (Figure 9.10) and dolomite (Figure 9.11), we will examine the cases where the porosity in the reservoir is 0.25, 0.15, and 0.05. The results indicate that the smaller the porosity the harder the reflection. The mineralogy is very important as well. For example, the Class I AVO reflection is very weak for the 25% porosity calcite reservoir while it is well pronounced for the dolomite reservoir at the same porosity. At the lowest porosity, 0.05, the wave hits a "brick wall" with a very strong positive reflection.

The next scenario is for tertiary limestone buildup constructed following Fournier and Borgomano (2007) and Grotsch and Mercadier (1999). We examine an interval that includes six lithofacies whose vertical order, thickness, porosity, density, and velocity are listed in Table 9.1. The resulting synthetic gather is shown in Figure 9.12. We observe a strong positive reflection at the first boundary, which is between the mudrock and packstone. The following reflections are relatively small until the boundary between the grainstone and quartz-rich packstone, where a strong positive reflection occurs followed by even stronger negative reflection at the interface between the quartz-rich packstone and relatively soft sandstone at the bottom.

This scenario is a simplification of a more complex real case described in Grotsch and Mercadier (1999) for the Tertiary Malampaya gas field offshore from the Philippines. Still, the two sharp reflections in Figure 9.12 match the real seismic profile shown by these authors for the gas well drilled close to the margins of the limestone buildup.

The final scenario is for Permian dolostone constructed following Lucia (2007) and designed for a West Texas oil field (Figure 9.13). The sequence includes three lithofacies: dolograinstone, dolopackstone, and dolowakestone. Their vertical arrangement and properties are listed in Table 9.2. In this case we observe two strong negative reflections, one at the interface between the upper dolowakestone and dolograinstone and the other at the same lithology contrast but located deeper in the interval.

10 Time lapse (4D) reservoir monitoring

10.1 Background

Time-lapse (4D) seismic methods attempt to quantify the difference in the seismic response of the subsurface before and after human interference, mainly hydrocarbon production and fluid injection. Multiple theories and case studies have been published. To mention just a few of them, we refer the reader to Calvert (2005), Osdal *et al.* (2006), Gommesen *et al.* (2007), Ebaid *et al.* (2008), Dvorkin (2008b), Ghaderi and Landrø (2009), Trani *et al.* (2011), and Ghosh and Sen (2012). The main idea is to interpret the temporal changes observed in the seismic response of the subsurface in terms of hydrocarbon saturation, pore pressure, and temperature in order to ascertain bypassed pockets of hydrocarbons and, eventually, increase the recovery factor.

Perhaps the first glimpse into the physics of time-lapse monitoring was in Nur (1969) where the ultrasonic-wave compressional velocity was measured on a granite sample with just 1.5% porosity as the water was draining from the originally fully saturated sample (Figure 10.1). Later, Wang (1988) discovered that the velocity in rock samples saturated with heavy oil strongly decreases when the samples are heated.

Practical seismic-based reservoir monitoring has to overcome a number of obstacles, one of which is cross-referencing the data obtained on the same subsurface volume but at different time. One needs to bear in mind that the vertical coordinate in a seismic volume is the two-way travel time (TWT) of the seismic wave rather than the physical depth. Because the speed of sound may change as the reservoir is produced, this is not an easy task. The physics of 4D seismic methods is not very simple either: some reservoirs may significantly change their porosity (compact) during production, while in some reservoirs the texture itself may vary if a reactive fluid (such as CO_2) is injected (Vanorio *et al.*, 2011). Here we will concentrate on the effects of two temporal variables, pore fluid and pore pressure, on the elastic properties of rock. These effects are dominant in most clastic reservoirs.

The theory of fluid substitution, that is, calculating the changes in a rock's elastic properties as the pore fluid changes, is well developed (Chapter 2). Nevertheless, there are choices to make: one must determine which type of fluid distribution, uniform or patchy, is most relevant to interpreting remote monitoring results. Knight *et al.* (1998)

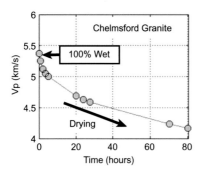

Figure 10.1 Velocity versus time in a 1.5% porosity Chelmsford granite sample (Nur, 1969).

provide a patchy-fluid-substitution theory using the capillary pressure equilibrium principle and, thus finding out how different parts of heterogeneous rock saturate during drainage or injection. Still, this theory contains a number of unknown variables, such as porosity and permeability heterogeneity. To bypass (but not eliminate) these complications, here we will use the principle of forward modeling "what-if" scenarios with the hope that if the modeled response matches the observation, the model describes the reality.

In spite of the multitude of published velocity versus stress experimental results, the effect of pore pressure on the elastic properties of rock still resists simple and basic theoretical generalization. Although the mechanism of these relations is (arguably) understood – as the pore pressure increases it opens up cracks in the rock and, thus softens the rock – it is unknown how many and what type of such defects emerge during injection.

To quantitatively ascertain the effects of pore pressure on the elastic properties we typically use laboratory experiments where the process is essentially reversed. Instead of varying the pore pressure at fixed confining stress, we keep the pore pressure constant and change the confining pressure. Because most of the laboratory velocity measurements are by necessity conducted at ultrasonic frequency, we need to be aware that the frequency difference between the laboratory (about 10^6 Hz), well measurements (10^3 to 10^4 Hz), and seismic data (10^1 to 10^2 Hz) usually make the laboratory data for liquid-saturated samples not quite applicable to seismic interpretation. This is why it is prudent to conduct the ultrasonic pulse transmission measurements on dry (air-saturated) samples and then conduct low-frequency fluid substitution on such data. Figure 10.2 shows some of such results obtained on sandstone samples. Notice that in both Fontainebleau sandstone and Ottawa sand, Poisson's ratio decreases with the decreasing confining pressure (Dvorkin *et al.*, 1999).

Figure 10.3 helps the reader to appreciate the diversity of velocity–pressure behavior in sandstone samples (data from Han, 1986): the maximum value reached at high pressure, the initial low-stress value, and the shape of the curves.

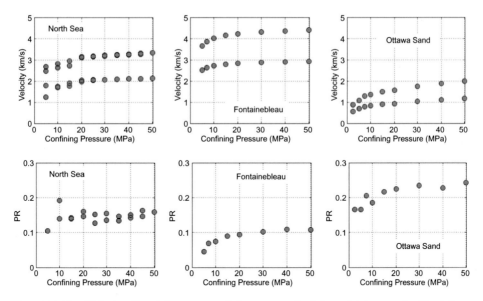

Figure 10.2 Top: Velocity (*P* and *S*) versus confining pressure in (left to right) room-dry North Sea sandstone (courtesy Statoil), Fontainebleau sandstone (Han, 1986), and Ottawa sand (Yin, 1992). Bottom: Poisson's ratio versus pressure for the same samples.

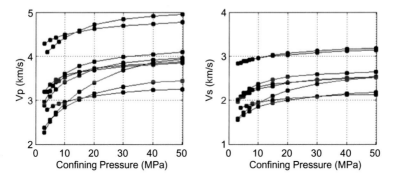

Figure 10.3 The *P*- and *S*-wave velocity versus confining pressure for a number of consolidated sandstone samples (Han, 1986).

A straightforward way of using laboratory dry-rock velocity versus confining pressure to assess the effect of changing pore pressure at constant overburden is to mirror-flip graphs such as those shown in Figures 10.2 and 10.3; the mirrored graphs are shown in Figure 10.4. In the mirrored graph, the measured dry-rock velocity is plotted versus the difference between 50 MPa, which is the maximum confining pressure in this experiment, and the varying (smaller) confining pressure, P_c. This difference is assumed to be the pore pressure, P_p, in the air-saturated sample. In other words, the velocity at zero P_p is the same as at $P_c = 50$ MPa and the velocity at $P_p = 30$ MPa and

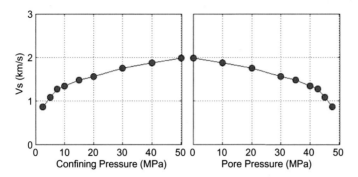

Figure 10.4 Dry-rock velocity measurement on Ottawa sand (Yin, 1992). Left: The original experimental data. Right: The same data but plotted versus the difference between the maximum confining pressure 50 MPa and varying confining pressure. This difference is assumed to represent the varying pore pressure.

confining pressure 50 MPa is the same as at confining pressure $P_c = 20$ MPa and zero pore pressure.

The operation illustrated by Figure 10.4 appears almost obvious. Yet, there is an important assumption behind this coordinate transform: we implicitly assumed that the velocity depends on the difference between the confining pressure and pore pressure, which we call the *differential pressure, P_d*:

$$P_d = P_c - P_p. \tag{10.1}$$

In other words, we assume that the elastic properties measured at, for example, 40 MPa confining pressure and 20 MPa pore pressure are the same as those measured at 30 MPa confining pressure and 10 MPa pore pressure.

Experiments and theories (see Mavko *et al.*, 2009) indicate that this statement is not absolutely correct. In fact, the elastic properties depend on the so-called *effective pressure, P_e*:

$$P_e = P_c - \alpha P_p, \tag{10.2}$$

where α is the effective pressure coefficient ($\alpha \leq 1$). It is very likely that Eq. (10.2) is an approximation and the functional form of the relations discussed here is more complicated.

Still, there is a problem with practical use even of the simple Eq. (10.2). The reason is that the coefficient α can be different for different rock properties (Mavko *et al.*, 2009). This is an example of a conflict between perfect science and practical applications: there are too many unknowns to practically implement the former. Hence, here we will settle for a "good enough" solution and assume that all the variables of interest depend on the differential pressure, $P_d = P_c - P_p$:

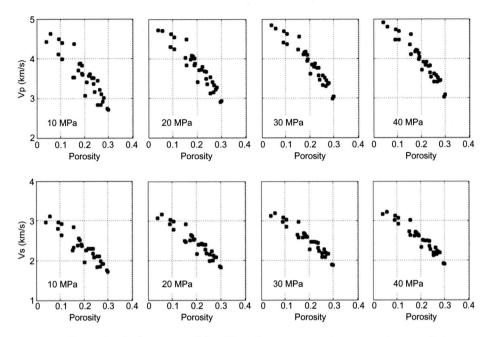

Figure 10.5 Dry-rock velocity on consolidated low-clay-content sandstone samples at confining pressure 10, 20, 30, and 40 MPa (Han, 1986).

$$V_p = F_1(P_c - P_p); \quad V_s = F_2(P_c - P_p), \tag{10.3}$$

where F_1 and F_2 are experimentally determined functions. Moreover, although the porosity usually varies with pressure, especially so in softer rock, let us neglect this variation and assume that the porosity and, as a result, the bulk density of a dry sample are both pressure independent.

It is important to mention here that the *velocity–pressure relations should not be combined with porosity–pressure relations to obtain velocity–porosity trends*. The latter are usually established at a fixed differential pressure and may vary with P_d (Figure 10.5).

10.2 Fluid substitution on velocity–pressure data

The bulk modulus and density of reservoir fluids change versus the pore pressure and temperature (Chapter 2). Here, for the sake of simplicity, we will neglect these variations and keep the fluid properties constant for brine salinity 80,000 ppm; oil API 25; gas gravity 0.75; gas-to-oil ratio 200; pore pressure 20 MPa; and temperature 70°C. The resulting bulk moduli and densities (Batzle and Wang, 1992) are listed in Table 10.1.

Table 10.1 *Fluid properties computed as explained in this section.*

Fluid	Bulk Modulus (GPa)	Density (g/cm³)
Gas	0.044	0.186
Oil	0.561	0.709
Brine	2.865	1.043

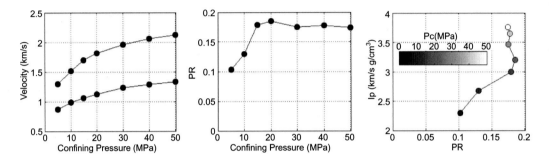

Figure 10.6 Dry-rock elastic properties of Ottawa sand from Han (1986). This sample is different from that used by Yin (1992) which is shown in Figure 10.4. From left to right: Velocity (*P* and *S*) versus confining pressure; Poisson's ratio versus confining pressure; and the *P*-wave impedance versus Poisson's ratio color-coded by the confining pressure, as shown in the color bar.

The sample used in this chapter is very soft Ottawa sand (Han, 1986). To simplify the following examples, we assume that its porosity, 0.33, is approximately pressure independent and so is the dry-rock bulk density, 1.77 g/cm³. The elastic properties of this sample as measured are shown in Figure 10.6.

Let us now conduct fluid substitution on these data by theoretically filling the sample with gas, oil, and water, all for 100% saturation. The resulting I_p and v are plotted versus the confining pressure in Figure 10.7. Assume next that this sample represents a reservoir located at certain depth where the overburden stress is 40 MPa and the original pore pressure is P_p = 20 MPa, so that the original (initial) differential pressure is 20 MPa. The only variable pressure in this example is P_p, which we will allow to vary from its initial value by ±10 MPa so that the respective differential pressure can vary between 10 MPa for P_p = 30 MPa and 30 MPa for P_p = 10 MPa. The respective impedance for the sample fully saturated with gas, oil, or water is shown in Figure 10.8. The plot on the right is convenient for predicting the impedance change due to various processes. For example, assume that we increase the pore pressure so that the free gas originally present in the oil/gas system goes into solution, leaving the reservoir without free gas. This hypothetical transition is shown by an arrow in Figure 10.8 (right). The net result is no change in the impedance, as the effects of changing the pore fluid and increasing the pore pressure counteract each other.

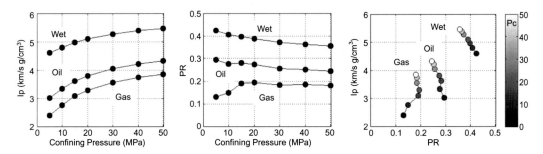

Figure 10.7 Same as Figure 10.6 (with the impedance shown instead of velocity) but for the sample theoretically saturated with gas, oil, and water, as marked on the plots.

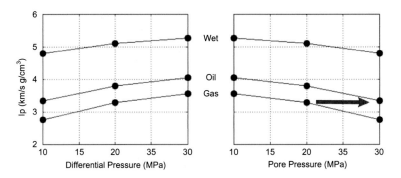

Figure 10.8 Impedance versus differential (left) and pore (right) pressure as explained in the text. The arrow in the plot on the right shows the hypothetical transition from gas to oil as the pore pressure increases and gas bubbles collapse (gas goes into solution).

Of course, the example of condition variation in the reservoir as illustrated in Figure 10.8 is not entirely realistic, as in a real hydrocarbon reservoir there is never 100% hydrocarbon saturation and whatever the initial saturation is, it cannot be reduced to zero due to residual hydrocarbon saturation. This simply means that we cannot transit from, for example, a gas reservoir with 20% water saturation and 80% gas saturation to that with 100% water saturation and zero gas: there will always be gas left in the pores.

Hence, to make the example more realistic and still simple, we will assume that the reservoir is a gas reservoir with the initial $S_w = 0.20$. This reservoir is gradually depleted to 20% gas saturation ($S_w = 0.80$). The irreducible water saturation $S_{wi} = 0.02$ and the residual gas saturation is also 0.02, meaning that the maximum possible water saturation is 0.98.

Let us first assume that as the gas is depleted and water saturation increases, the resulting two-phase fluid distribution is uniform and, hence, the effective bulk modulus of the pore fluid is the harmonic average of those of gas and water (Chapter 2). The resulting I_p and v are plotted versus S_w and also versus each other in Figure 10.9.

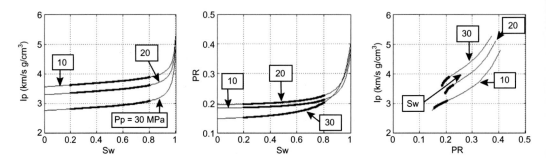

Figure 10.9 Impedance (left) and Poisson's ratio (middle) versus water saturation using the assumption that fluid distribution is uniform. The third graph shows a cross-plot of impedance versus Poisson's ratio as water saturation increases (arrow). Each of the three curves is for a fixed pore pressure as marked in the plots. Thin curves are for water saturation varying from zero to 100% while bold curves are for water saturation ranging from 20 to 80%.

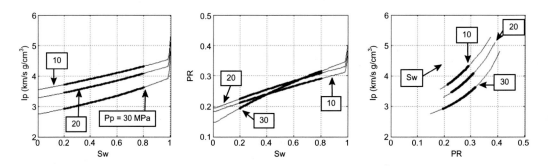

Figure 10.10 Same as Figure 10.9 but for the patchy saturation scenario.

There is very little variation of either impedance or Poisson's ratio versus water saturation in the range $0.20 < S_w < 0.80$, a fact also discussed in Chapter 2. Yet, published field studies indicate that there are noticeable elastic-property changes during production.

This apparent seismic visibility of saturation changes makes us hypothesize that the fluid distribution pattern due to the transient process of production may be patchy. The respective theoretical curves are plotted in Figure 10.10. Now we observe discernible variations in the rock's elastic properties with respect to water saturation.

10.3 Synthetic seismic gathers

To illustrate the effect of varying pressure and saturation on seismic data we will examine two production scenarios: (A) the gas reservoir is depleted from its initial state (P_p = 20 MPa and S_w = 0.20) with pore pressure dropping from 20 to 10 MPa

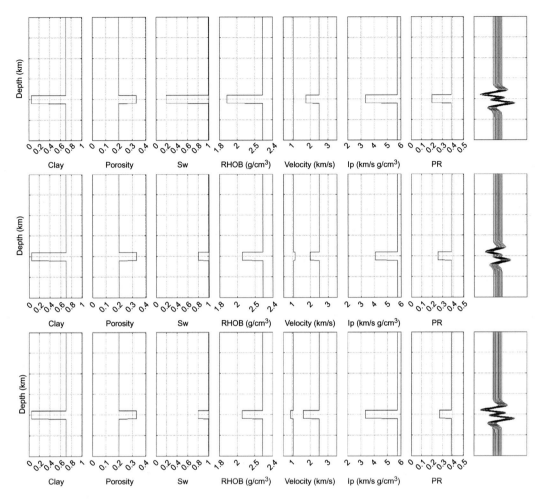

Figure 10.11 Depth curves and synthetic seismic gathers for the initial state of the reservoir (top); Scenario A (middle); and Scenario B (bottom). The vertical distance between the adjacent horizontal grid lines is 50 m.

and water saturation increasing from 0.20 to 0.80 and (B) the reservoir is under water injection with S_w varying in the same range but the pore pressure increasing from 20 to 30 MPa. The constant background is soft wet shale with 70% clay content and 20% porosity.

The initial state as well as Scenarios A and B depth curves and synthetic seismic gathers are shown in Figure 10.11. Synthetic gathers were generated using a 40 Hz Ricker wavelet. The three resulting gathers are compared to each other in Figure 10.12.

Because the impedance in the gas sand is much smaller than that in the shale, the amplitude is negative at the shale/sand interface. The amplitude's absolute value decreases from the initial state to Scenario A. The primary reason is the stiffening of the gas sand by the differential pressure increasing from 20 MPa at the initial state to

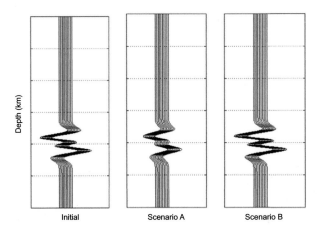

Figure 10.12 Synthetic gathers from Figure 10.11 for the initial state (left); Scenario A (center); and Scenario B (right). The vertical distance between the adjacent horizontal grid lines is 50 m.

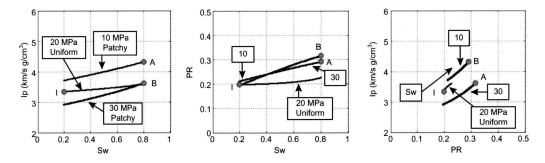

Figure 10.13 Same as Figure 10.10 but for the 0.20 to 0.80 saturation range and uniform saturation and 20 MPa pore pressure; patchy saturation and 10 MPa pore pressure; and patchy saturation and 30 MPa pore pressure. Symbols marked "I," "A," and "B" are for the initial state of the reservoir, Scenario A, and Scenario B, respectively. The diagonally pointing arrow in the third plot shows the direction of increasing saturation.

30 MPa for Scenario A. Increasing saturation in the patchy mode adds to this pressure-driven impedance increase.

In contrast, there is little change in the amplitude between the initial state and Scenario B. Here, the effect of increasing saturation acting to increase the impedance is counteracted by the differential pressure decreasing from 20 MPa at the initial state to 10 MPa and acting to reduce the dry-frame stiffness.

To further explain these results, refer to Figure 10.13 with the display analogous to those in Figures 10.9 and 10.10. In this figure we plot the elastic property variations in the water saturation range between 20 and 80% for uniform saturation and $P_p = 20$ MPa; patchy saturation and $P_p = 10$ MPa; and patchy saturation and $P_p = 30$ MPa. The beginning of the first curve corresponds to the initial state of the reservoir; the end of

Table 10.2 *Elastic properties and densities of the strata used in 4D forward modeling.*

Stratum	V_p (km/s)	V_s (km/s)	Density (g/cm³)
Shale	2.507	1.078	2.289
Reservoir Initial	1.773	1.089	1.886
Reservoir Scenario A	1.991	1.141	2.056
Reservoir Scenario B	1.631	0.914	2.056

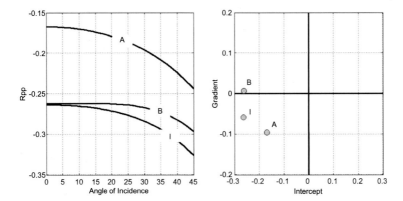

Figure 10.14 Left: AVO curves at the upper shale/sand interface for the initial state of the reservoir as well as for production scenarios A and B (as marked in the plot). Right: Gradient versus intercept for the three states of the reservoir.

the second curve corresponds to Scenario A; and the end of the third curve corresponds to Scenario B. The respective I_p and v changes explain the synthetic seismic results.

Finally, let us produce AVO curves at the shale/sand interface for all three cases examined in Figures 10.11 to 10.13. The velocities and bulk densities above and below this interface for the initial state as well as the two scenarios examined here are listed in Table 10.2. The respective AVO curves as well as the intercept–gradient cross-plot are shown in Figure 10.14.

Notice that for Scenario B the computed gradient is slightly positive. This is because the AVO curves were calculated using the Zoeppritz equations (Chapter 4) while the gradient was computed from these curves assuming that the *P*-to-*P* reflectivity is proportional to the sine square of the incident angle which, in this case is an approximation.

10.4 Conclusion

The main purpose of time-lapse reservoir monitoring is to determine the fluid saturation during production in space and time in order to better plan the recovery and optimize the placement of additional wells. The interpretation of field observations is based on the assumptions about fluid distribution pattern as well as pore pressure variation. As for any seismic data interpretation, this process is not unique. One way of dealing with this uncertainty is to conduct a number of forward-modeling scenarios to find a match with the observed seismic signatures. Such modeling may help bracket the interpretation results as well as eliminate unlikely outcomes.

In the example discussed in this chapter, we only examined two input variables, saturation and pressure. Depending on the reservoir type, additional variables have to be taken into account, such as temperature in heavy oil sands or porosity variations in compacting chalk reservoirs. Rock physics theory provides us with the models to take such additional variables into consideration.

Here we examined the effects of production on the elastic properties of the subsurface. Field studies have shown that in some cases attenuation may also be a useful attribute for production monitoring (e.g., Eastwood *et al.*, 1994). Chapter 15 provides the reader with quantitative theories to include attenuation in forward modeling.

In practical 4D seismic applications, the scenarios presented here (and more) should serve as a feasibility study to be conducted before acquiring 4D data to help decide whether and how such seismic data should be acquired.

Part IV

Frontier exploration

11 Rock-physics-based workflow in oil and gas exploration

11.1 Introduction

A key challenge faced by geoscientists is to establish and validate the economic potential of a hydrocarbon exploration play using seismic amplitudes. This validation requires the development and implementation of interpretation techniques for calibrating rock properties to seismic data. Quantitative seismic interpretation techniques have been applied with frequent success to predict lithology and fluids in areas with extensive local well control (Avseth *et al.*, 2005). The application of the same techniques is problematic in frontier basins where the nearest well control is some distance away. The main reason is that when interpretation techniques are extrapolated outside their original range of calibration, seismic anomalies cannot be reliably recognized, predicted, and validated.

Validation of recognized seismic anomalies comprises two stages (a) rock-physics-based modeling for amplitude calibration described in this chapter, followed by (b) detailed amplitude interpretation and risk analysis (Chapter 12). This integrated methodology for regional exploration combines rock physics modeling and seismic-based evaluation techniques, allowing the seismic interpreter to predict, quantify, and extrapolate seismic response changes by linking them to rock-property variations and plausible geologic scenarios. This method includes examination of depositional and diagenetic processes to understand the effects of geology on seismic responses as well as possible differences between the interpretation model at the current prospect and that established elsewhere using well control and other known inputs.

Figure 11.1 shows a detail of a generalized workflow that can be used to interpret seismic amplitude anomalies in terms of lithology, fluid content, porosity, rock texture, and pore pressure, focusing on establishing a rock-property-based seismic interpretation framework, generating a catalogue of seismic responses of potential exploration success and failure scenarios, and, thus de-risking an exploration opportunity. This workflow includes several key components: (a) identifying elastic rock types, such as shale and wet, oil, and gas sand; (b) forward modeling geologic scenarios (in a way shown throughout this book); (c) generating synthetic seismic gathers for these

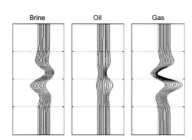

Figure 11.1 A detail of interpretation workflow: synthetic scenario-based seismic gathers to be compared to the actual seismic anomaly in real data cube. In the real-seismic display, red is for negative while blue is for positive amplitude. For color version, see plates section.

scenarios; and (d) comparing these synthetic seismic responses to the seismic amplitude anomaly under examination.

11.2 Rock physics modeling for amplitude calibration

Seismic responses related to oil and gas accumulations are called *direct hydrocarbon indicators* or DHIs. After recognizing key amplitude anomalies in seismic data and before establishing associated DHIs, it is critical to understand the possible changes in rock properties that could be associated with the seismic anomaly and to find out if the available seismic field data can be used to predict the expected lateral changes of the relevant rock properties at the seismic scale (Chapter 12). In general, the predictive use of seismic data is a direct function of the formation impedance contrast versus depth, the lateral variability of the reservoir and associated bounding rocks, and, obviously, the quality of the seismic data.

First, relevant depth trends of the rock's elastic properties are used to examine whether the expected variation of rock properties can produce a clear effect on seismic data at the prospect location, assuming that the seismic data are of ideal quality. Later in the workflow, rock-physics-based seismic forward modeling is used to investigate whether the seismic field data can predict the key rock property spatial variations. This procedure consists of the following steps:

- Log and seismic data quality control and conditioning, including repair of poor-quality logs using rock physics transforms.
- Time–depth calibrations. Analysis of check-shot velocity surveys and calibration of well and seismic processing velocities.

- Generation of synthetic seismograms for key wells.
- Rock typing and upscaling.
- Rock property trend analysis, rock physics diagnostic and model formulation. Integration with existing thermal, burial, and reservoir quality prediction models, based on regional basin modeling and petrographic analogue data. Fluid elastic properties modeling for the expected pressures and temperatures in the prospect area.
- Use of predictive models at the prospect location, extrapolated through velocity-based vertical effective stress (VES) and reservoir porosity predictions. Scenario-based AVO forward modeling to determine the seismic response of the assumed reservoir/seal pairs and investigating the DHI types and their applicability at the site.

11.3 Log and seismic quality control and conditioning

Quantitative seismic interpretation is built upon reliable seismic and borehole data. The required input dataset generally includes pre-stack and stacked seismic volumes, processing velocity models, well logs, and geological data (e.g., lithology description and petrophysical core analysis). It is important to verify whether the log data quality is high enough and the available geological and petrophysical data are sufficient for regional calibration.

It is critical that the key log datasets are carefully quality controlled and selected for rock physics analysis. Wells are selected based on the log availability and vertical and horizontal sampling of the key geological units. A regular geographical and geological sampling that captures the rock property variability is desirable.

The main focus of this initial task is to remove any spurious log data, especially density and compressional- and shear-wave velocity in the entire well, and to only keep reliable information, free from undesirable borehole and acquisition effects and errors that otherwise can propagate throughout the data analysis and can compromise the final rock physics models. Common sources of errors in log data include depth mismatch, borehole wall rugosity due to washouts, mud filtrate invasion, and mudrock (shale) alteration. Log curves are analyzed and edited for unreliable segments using a range of manual and empirical techniques. Extensive quality checks of the logs must be conducted to address cycle skipping and noise spikes in the sonic logs, environmental effects associated with a caved or rough hole in the density logs, and erroneous first arrival picking in the shear velocity log usually associated with very slow rocks.

Comparison of well data with existing rock physics models as well as with relevant empirical regressions is included in log edition. Once such models and regressions are calibrated to become site-specific, low-quality log curves are repaired and missing logged intervals are re-generated as inputs to seismic forward modeling.

Seismic data must have sufficient quality to make reliable interpretations and valid conclusions. The amplitude and phase preservation of both stacked and gather data are key to successful quantitative interpretation. Communication between the seismic processor and the quantitative interpreter is critical to ensure that the seismic processing has preserved the relative AVO response. The quality of pre-stack data for AVO analysis is crucial, as the key seismic attributes, such as intercept and gradient are not robust in the presence of small normal move out (NMO) errors. To avoid unreliable estimates of AVO attributes, the priority is to have available migrated gathers showing very good event time alignment with offset and preserved amplitudes (Yilmaz, 2001).

11.4   Velocity in exploration seismology

There are two main types of velocities in exploration seismology: physical velocities and velocity measures. Physical velocities are the speeds of elastic waves in earth, including instantaneous velocity, compressional and shear velocities, and phase and group velocities (Mavko *et al.* 2009). In contrast, velocity measures are estimates from the seismic data analysis that have the same units as the physical velocities but are only related in some indirect way. Good examples of velocity measures include interval velocity, stacking velocity, migration velocity, and average, mean, and root-mean-square (RMS) velocities (Yilmaz, 2001).

In situ velocity measurements need to use a borehole and include two kinds of direct methods or well surveys: sonic logging and check-shot surveys. Sonic logging displays the travel time of compressional and shear waves versus depth. The wireline sonic tool emits a sound wave that travels from the source through the formation and back to a receiver. Qualitatively, the sonic log is used in geological correlations and the identification of lithology, source rocks, and normal compaction and overpressured zones (Rider, 2002). Traditional quantitative applications include porosity evaluation in consolidated sandstones. Although sonic logging is often considered a tool to be used by petrophysicists and geologists, this log was originally invented as an aid to seismic exploration to provide interval velocity profiles for time–depth calibrations and the generation of synthetic seismograms (Sheriff and Geldart, 1995).

Modern sonic tools can record full waveforms that include the *P*-wave, *S*-wave, and the Stoneley wave arrivals. These in situ elastic velocity measurements are vital in seismic interpretation studies, including lithology, porosity, and hydrocarbon predictions. Sonic logging tools can often provide reliable values for velocities through the well casing. It is a desirable feature to be used in soft shales which may become mechanically unstable in an open hole (Paillet *et al.*, 1992).

Check-shot surveys measure the seismic travel time from the surface to a station in depth. The *P*-wave velocity can be measured directly by lowering a geophone in the well, sending acoustic energy from the surface, and recording the first arrival of the wave. Apart from measuring the first arrival, the check-shot survey can also be used to analyze the full wave train. This so-called vertical seismic profiling (VSP) has to be analyzed in terms of up- and down-going seismic events with the results displayed in a seismic image format (Hardage, 1985). VSP uses numerous geophones positioned closely at regularly spaced intervals in the wellbore, as opposed to irregularly positioned geophones in check-shot surveys. VSP data, when properly recorded and processed, provide the best data for detailed seismic event identification and wavelet determination. VSP is also a key technique to generate a depth-to-travel-time calibration (Hardage *et al.*, 1994).

Seismic data provide a structural image of the reservoir zone that depends on lithology, porosity, type of fluid, and temperature and pressure. To predict these physical properties and conditions, it is necessary to establish relations between wave propagation attributes (i.e., velocity and impedance) and petrophysical and lithological properties. As discussed before, such relations use controlled experiments, including well data and laboratory measurements.

11.5   Time-to-depth calibration

The example below deals with an offshore prospect called the offshore area (OA) and includes seismic data and also well data from four offset wells outside the prospect. Sonic logs can be displayed in different ways to be analyzed and integrated with the seismic data. Two important displays are the interval velocity and vertical travel time versus depth (Figure 11.2). The interval velocity is usually estimated from the reciprocal of the sonic interval transit time, or from the stacking velocities following the Dix equation (Sheriff and Geldart, 1995). To compare sonic to seismic data, sonic velocity logs must be smoothed (upscaled to the seismic scale) using the Backus average (Mavko *et al.*, 2009).

The sonic log usually samples one reading per half-foot (about 15 cm); in contrast the check-shot data sampling is a sample per 50 to 200 m. The sonic log velocity information is used to interpolate between such discrete measurements. Sonic transit time data frequently differ from check-shot data (Box and Lowrey, 2003). The difference between the two-way time estimated from the discrete check-shots and the sonic log is called velocity drift. Sometimes the integrated sonic times are longer than the check-shot times and the drift is considered negative and is usually caused by borehole formation damage during drilling as well as cycle skips in sonic waveform processing. In contrast, positive drift may be related to frequency effects (Marion *et al.*, 1994), where lower frequency seismic waves (10 to 90 Hz) travel slower than higher frequency sonic waves (20 to 30 KHz). This difference may be also caused by mud filtrate invasion into the formation sampled by the sonic tool: partial replacement of compressible hydrocarbon in the pores with less compressible water can make the rock stiffer and faster (see Chapters 2 and 10 for fluid substitution and discussion of patchy saturation developing during transient fluid flow).

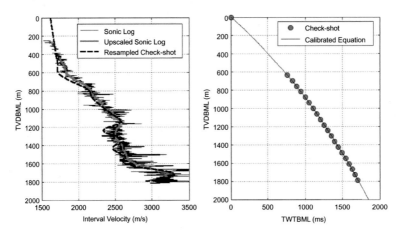

Figure 11.2 Left: To compare the sonic data to seismic data, sonic velocity logs must be brought to the same scale using the Backus average. Right: Time–depth calibration example using check-shot data.

To locate the reservoir in space, the seismic response needs to be converted from travel time to depth. A wide variety of functional forms are available to estimate depth from seismic velocity and two-way travel time. The most popular methods are the two-parameter forms including linear (Slotnick, 1936) and power-law equations (Faust, 1951, 1953; Evejen, 1967). In practice, functions or trends, relating velocity and travel time to depth are calibrated using check-shot and VSP surveys and sonic velocity logs. The earliest, simplest, and most widely used analytic expression to describe the variation of instantaneous velocity (V_{ins}) with depth (z) is due to Slotnick (1936):

$$V_{ins} = V_0 + kz, \tag{11.1}$$

where V_{ins} is defined as the speed of the wave front along the direction of its propagation. The constants in this equation are established using sonic data (Marsden *et al.*, 1995).

A similar linear expression is often used to describe the velocity of a layer in the subsurface or interval velocity (V_{int}) as a function of depth:

$$V_{int} = V_0 + kz, \tag{11.2}$$

where V_{int} is the average velocity over a continuous interval in the subsurface.

If the velocity is a linear function of depth, it is exponential with respect to the vertical travel time (Slotnick, 1936):

$$z(t) = \frac{V_0}{k}(e^{kt/2000} - 1), \tag{11.3}$$

where the travel time, t, is in milliseconds, velocity is in m/s, and depth is in m. The coefficient 2000 is used to translate the two-way travel time in milliseconds into one-way travel time in seconds. An example of using this equation for relating depth to time is shown in Figure 11.2 (right).

An equation reciprocal to Eq. (11.3) converts depth to time as

$$t(z) = \frac{2000}{k} \ln\left(1 + \frac{kz}{V_0}\right). \tag{11.4}$$

These functions can be calibrated using sonic log and check-shot data from multiple wells and generally describe the dominant overburden effects. Check-shots provide the most accurate time-to-depth information. Two methods are used to estimate V_0 and k: (a) conventional cross-plot between midpoint depth and interval velocity in the linear equation and (b) calibration of the exponential equation by minimizing the difference between model-predicted and actual data using mathematical optimization. Both constants, V_0 and k, are affected by lithology, tectonic uplift, geopressure, and rate of deposition (Japsen, 1993, 1998; Marsden *et al.*, 1995; Japsen *et al.*, 2007).

11.6 Rock typing

Accurate lithological identification of reservoir and seal rocks is required for reliable rock physics modeling. Lithology flags must be generated for the rock property analysis indicating the key rock types. Well-based lithology information usually comes from mud logging, direct physical sampling from cores, and interpretation from wireline logs.

The initial coarse lithology description given by mud logs must be verified by well log GR, SP, resistivity, sonic, density, and neutron data. Additional inspection of log data helps establish trends, baselines, and absolute values of rock properties (Rider, 2002). Log-based lithology identification has to be compared to core data. If an inconsistency is found, both data sources have to be re-examined. Special attention must be paid to identify lithologies like coals, volcanic ashes, marls, and calcite stringers, which can be associated with misleading seismic response interpretations.

11.7 Seismic forward modeling

As discussed throughout this book, to identify the seismic response of key reservoir rocks and associated stratigraphic units, well data are related to seismic data via synthetic seismograms. A synthetic seismic trace is generated by convolving the reflectivity series derived from sonic and density logs with the wavelet derived from seismic data (Waters, 1992). The sonic log is usually calibrated by the check-shot survey before combining it with the density log to produce the acoustic impedance (Box and Lowrey, 2003). The often-observed mismatch between the synthetic and real seismic data has given rise to a sub-discipline known as well-to-seismic tie. This field is fairly nebulous and requires as much art and experience as it requires logic. It is important to bear in mind that no data contain the ground truth: many factors may affect the veracity of well and seismic data as well as the process of synthetic seismic generation that requires an appropriate wavelet that is impossible to reliably establish as it varies with depth and location. The logic used in well-to-seismic tie has to be as simple as possible and rely on circumstantial evidence. For example, if VSP confirms a synthetic seismogram while seismic data do not, it may be prudent to rely on the synthetic data and reprocess the seismic data or start looking for indications of synthetic-seismic events in less-than-perfect real seismic volumes.

The potential of using AVO for enhancing DHIs is established through forward modeling of the seismic response for multiple geological scenarios. The goal is to generate a catalogue of seismic responses of potential exploration success and failure. Realistic scenarios are built by modeling individual rock types and associations of rock types in accordance with the laws of geology and local stratigraphy. Sometimes amplitude anomalies are not present in the field seismic data. In such cases, special attention has to be paid to other indicators, such as flat spots that may indicate fluid contacts or diagenetic transitions, such as between different types of opal.

11.8 Rock property upscaling

The seismic velocity in stratified media depends not only on the rock and fluid properties, but also on the scale of the seismic data compared to the scale of the sedimentary section. Velocities are controlled by the ratio of the wavelength, λ, to the layer thickness, d, and the apparent velocity may be higher where λ/d is much smaller than 1, compared to the opposite case. Where $\lambda/d \ll 1$, the wave propagates as a ray and the total travel time through a sequence is exactly the sum of the travel times through individual layers. In the opposite case where the layer thickness is much smaller than the wavelength, the wave averages the elastic compliances of individual layers and the apparent velocity is decreased (see discussion in Marion *et al.*, 1994 and Mavko *et al.*, 2009).

Ideally, predictive rock physics models should be generated at the seismic scale. A log-scale relation should not be unconditionally applied to seismic-scale data, because an elastic property at a point in space cannot be accurately recovered from an experiment that employs large wavelengths (Dvorkin *et al.*, 2003; Dvorkin and Uden, 2006). There are no universal rules (except for forward seismic modeling) for translating the core-scale ($\lambda \approx 0.005$ m) and log-scale ($\lambda \approx 0.5$ m) measurements into the seismic scale

($\lambda \approx 50$ m). Yet, there are indications that rock physics transforms may be approximately scale-independent (see Chapter 17 as well as Dvorkin, 2008c, 2009; Dvorkin and Nur, 2009; Dvorkin *et al.*, 2009; and Dvorkin and Derzhi, 2013).

Still, simple averaging rules can be used to upscale log data. Volume and mass properties can be upscaled using a running arithmetic averaging window, such as

$$\langle \rho \rangle = \frac{1}{n} \sum_{i=1}^{n} \rho_i, \tag{11.5}$$

where ρ_i are individual consecutive bulk density readings, $\langle \rho \rangle$ is the upscaled density, and n is the size of the averaging window. The elastic moduli can be upscaled using a running harmonic (Backus) averaging window (Mavko *et al.*, 2009):

$$\langle M \rangle = \left(\frac{1}{n} \sum_{i=1}^{n} \frac{1}{M_i} \right)^{-1}, \tag{11.6}$$

where M_i are individual consecutive elastic modulus readings and $\langle M \rangle$ is the upscaled elastic modulus. The length of the averaging window should be comparable with the wavelength, λ, for example, $\lambda/10$ (Sams and Williamson, 1993 ; Menezes and Gosselin, 2006) or even $\lambda/20$ (Rio *et al.*, 1996).

The depth curves thus upscaled are then divided into layers to identify "endpoint" lithology intervals, such as "pure" shale or "pure" sandstone. A measure of "purity" is the coefficient of variation for a property defined as the ratio of the standard deviation to the mean within a selected interval, also called the coefficient of heterogeneity. Once "pure" lithology intervals are defined, relations between rock properties are established for these lithology members by, for example, curve fitting using upscaled curves.

11.9 Depth trends, rock physics diagnostics, and model formulation

An important part of designing a site-specific rock physics transform is to estimate pressure and temperature depth trends and then use them to compute the bulk modulus and density of the reservoir fluids (Chapter 2). Figure 11.3 shows an example of pressure and temperature depth trends at calibration wells nearest to the prospect.

Pore pressure can be predicted prior to drilling in a number of ways, including using seismic interval velocity and basin modeling. Because geopressure impacts well design and production scenarios as well as hydrocarbon trap integrity, the prospect and play appraisal and economics highly depend on geopressure prediction. The main principle of pressure prediction is that seismic velocity increases with increasing differential stress.

There are various reasons for overpressure development. One is sediment compaction in geologic time as it is gradually buried. Sediment becomes overpressured if

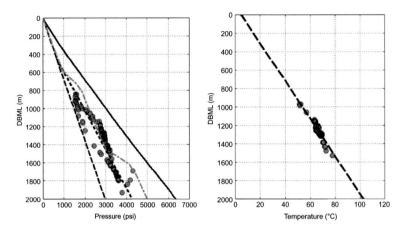

Figure 11.3 Left: Pressure and stress versus the depth below the mudline (DBML). Solid black line is the overburden stress; dashed black line is hydrostatic pressure; gray symbols are for pore pressure measured in calibration wells closest to the prospect; short-dash curve is the regional pore pressure trend; and dash-dot curve is for pore pressure from basin modeling. Right: Temperature versus depth as measured in the well (symbols) and according to the regional geothermal gradient (dashed line).

water cannot escape it while the overburden keeps increasing. In this case, the porosity and density remain large at depth, making the respective seismic velocity smaller than in the surrounding normally pressured strata. Another mechanism is tectonic uplift, where a reservoir originally residing at large depth is brought to a shallower subsurface. Its pore pressure, which is normal at large depth, becomes abnormally high at a smaller depth. In this case, the density remains approximately the same as at depth but the velocity becomes smaller as decreasing differential pressure opens thin microscopic cracks in the rock fabric.

In practical applications, only empirical locally calibrated equations are used to relate the interval velocity to pressure. These equations are discussed in, for example, Eaton (1975), Bowers (1995, 2002), Dutta (1987), Katahara (2003), and Gutierrez *et al.*, (2006).

Velocity and density depth trends are best calibrated by relevant well data upscaled according to Eqs (11.5) and (11.6) (Figure 11.4). Generally, mudrock's elastic properties are estimated using empirical depth trends as well as empirical trends between the actual velocity and seismic interval velocity, V_{int}. A simple velocity depth trend for mudrock can be linear versus depth, z:

$$V_{pMud} = a + bz. \tag{11.7}$$

Mudrock velocity can be also linearly related to the interval velocity as

$$V_{pMud} = c + dV_{int}, \tag{11.8}$$

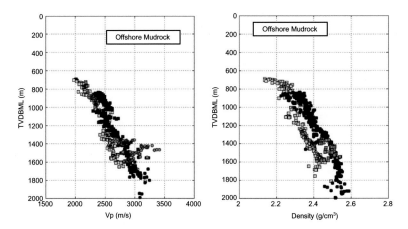

Figure 11.4 Upscaled velocity (left) and density (right) versus depth for four offshore wells (different symbols for different wells).

where the constants in Eq. (11.8) are of course different from those in Eq. (11.7).

Yet another way of determining variations of V_{pMud} with depth is by relating it to both depth and interval velocity as

$$V_{pMud} = A + BV_{int} + Cz, \tag{11.9}$$

where, once again, the constants are different from those used in the previous two equations.

The next task is to establish depth trends for the S-wave velocity, V_s. Many V_s predictors are discussed in Chapter 2. Here we introduce yet another approach proven to be useful in offshore basins. First we use a power-law equation relating the Poisson's ratio of mudrock, v_{Mud}, to its P-wave velocity, V_{pMud}, as

$$v_{Mud} = 0.5 - a(V_{pMud} - 1500)^b, \tag{11.10}$$

where the constants a and b are, of course, different from the constants used in previous equations and the units of the velocity are m/s. This equation is designed to satisfy the conditions at the mudline where rock is suspension, its velocity is approximately 1500 m/s and Poisson's ratio is 0.5. Once the coefficients in Eq. (11.10) are locally calibrated, we use the standard theory of elasticity equation to find V_{sMud} as

$$V_{sMud} = V_{pMud} \sqrt{\frac{1 - 2v_{Mud}}{2(1 - v_{Mud})}}. \tag{11.11}$$

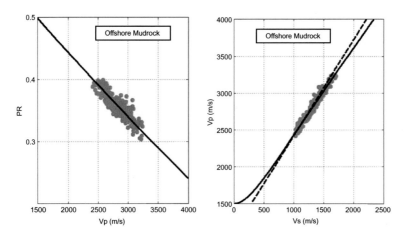

Figure 11.5 Left: Poisson's ratio according to Eq. (11.10) (black curve) fitted to local upscaled data (gray symbols). Right: The same but plotting the *P*- versus *S*-wave velocity. Black curve is according to Eqs. (11.10) and (11.11) while the dashed curve is according to Greenberg and Castagna (1992).

An example of calibrating Eq. (11.10) to local upscaled data is given in Figure 11.5 with $a = 0.00014$ and $b = 0.96$.

A similar approach is used to relate the bulk density, ρ_{Mud}, to the *P*-wave velocity, V_{pMud}, in mudrock. The functional form used is

$$\rho_{Mud} = 1.32 + a(V_{pMud} - 1500)^b, \tag{11.12}$$

where the constants a and b are, once again, different from the constants used in previous equations; the units of the velocity are m/s; and the density is in g/cm³. Figure 11.6 (left) shows data for the offshore area (OA) under examination and the fitting curve according to this equation with $a = 0.1575$ and $b = 0.275$. In Figure 11.6 (right), this curve is compared to other relations. It differs from these relations, which once again emphasizes the need for local-data calibration.

The next task after establishing the trends for mudrock is to do the same for the sand reservoir. Reservoir elastic properties, in particular the compressional velocity in sands, can be calculated using velocity–porosity transforms (Chapter 2). The porosity itself can be estimated from the so-called reservoir quality models. These models fall into three categories (Lander and Walderhaug, 1999): (a) effect-oriented models, such as statistical correlations of porosity and other variables (e.g., depth, temperature, and vertical effective stress); (b) process-oriented models, such as geochemical reaction-path models that are based on thermodynamics and kinetics principles; and (c) models empirically calibrated to geological datasets. Such statistical models for porosity loss are produced by fitting empirical functions to experimental core data or well log-based

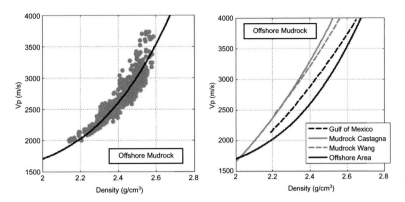

Figure 11.6 Left: Calibration velocity–density data and best-fit curve according to Eq. (11.12) with coefficients listed in the text. Right: Same best-fit curve (black) compared to an empirical curve for the Gulf of Mexico as well as empirical equations by Castagna *et al.* (1993) and Wang (2000).

datasets. These relations are strictly empirical and, hence, the constants must be calibrated locally.

Commonly used forms for the porosity–depth relation are linear, exponential, and power-law. The simplest relation between porosity, ϕ, and depth, z, is linear:

$$\phi = \phi_0 - cz, \tag{11.13}$$

where c is a positive constant.

This equation may predict negative porosity below a certain depth and is unable to adequately describe the rapid initial decline of porosity observed in clastic deposits (Giles, 1997). In a large depth range, the form of the porosity–depth behavior is better described by an exponential equation first proposed by Athy (1930) for shales:

$$\phi = \phi_0 e^{-cz}. \tag{11.14}$$

A power law of porosity reduction was advocated by, among others, Baldwin and Butler (1985).

$$\phi = 1 - (z / z_{max})^c, \tag{11.15}$$

where the constant, c, is naturally different for different equations.

Figure 11.7 shows the sandstone porosity versus depth curve established for OA using regional and local data with $\phi_0 = 0.40$ and $c = 0.0003$ m^{-1}.

A note on other model types. Process-oriented models, such as geochemical reaction-path methods, provide key insights into the diagenetic process but they are not well suited for practical applications as they require too many generally unknown input parameters. In contrast, the commercial versions of empirically calibrated models (e.g.,

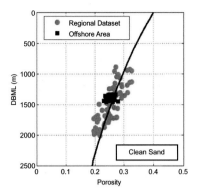

Figure 11.7 Regional data (gray) and curve from Eq. (11.14) with added local data for OA (black squares).

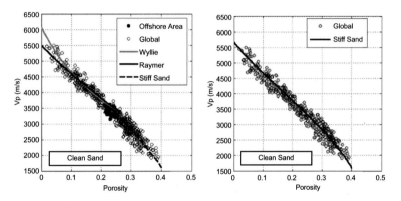

Figure 11.8 Global wet sandstone data (gray) and OA data (black). The model curves are listed in legends and explained in the text.

Touchstone™) are popular as they provide accurate predictions for quartz-rich sandstone using available geological inputs, such as the effective stress and temperature histories from basin modeling, together with the composition and texture from point count analysis of analogue thin sections (Lander and Walderhaug, 1999).

Three principal steps are used in reservoir rock modeling: rock physics diagnostics; model generation; and lithology and fluid perturbation. Rock physics diagnostics (Chapter 3) is a deterministic process. We prefer it to purely empirical local velocity–porosity fitting. Still, a rock physics model can be either empirical or based on effective-medium theoretical modeling. The key requirement for the model is to be predictive and geologically relevant.

Examples are shown in Figure 11.8 (left), where V_p is plotted versus porosity for a global dataset of upscaled well log data in very well-sorted wet clean (clay content below 10%) sandstone intervals. The model curves superimposed upon the data are from (a) Wyllie's time average (Mavko *et al.*, 2009): $V_p^{-1} = (1 - \phi)V_{ps}^{-1} + \phi V_{pf}^{-1}$, where

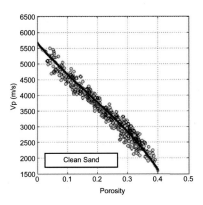

Figure 11.9 Same as Figure 11.8 but with the Eq. (11.16) curve. The stiff-sand model curve is hidden behind this best-fit curve.

the velocity in the solid phase, V_{ps}, was fixed at 6050 m/s and that in the fluid phase, V_{pf}, was 1500 m/s; (b) the Raymer–Hunt–Gardner (RHG) model with V_{ps} = 5484 m/s and V_{pf} = 1500 m/s; and (c) the stiff-sand model (for porosity greater than 0.25). In Figure 11.8 (right), we display only the stiff-sand model curve, this time computed for rock with 90% quartz and 10% clay and critical porosity 0.40. The bulk and shear moduli of quartz in this example are 37 and 45 GPa, respectively; these values for clay were 24 and 8 GPa. The density of quartz was 2.65 g/cm³, and the density of clay in this case was 2.71 g/cm³ to account for various heavy minerals present. This plot indicates that the stiff-sand model curve produced with these inputs matches the global data.

To further simplify practical use, we fit the data with a polynomial equation

$$V_p = 5669.2 - 11171\phi + 14664\phi^2 - 29828\phi^3, \tag{11.16}$$

where the velocity is, once again, in m/s. This curve is superimposed upon the data in Figure 11.9. This curve exactly matches the stiff-sand model and accurately describes the data.

The V_s prediction for sandstone follows the same scheme as described earlier for the mudrock (shale). First, Poisson's ratio is found using the equation

$$v = 0.5 - a(V_p - 1606.9)^b, \tag{11.17}$$

where $a = 1.0034 \times 10^{-4}$ and $b = 1.0152$. Finally, V_s is calculated from Eq. (11.11).

As usual, the bulk density in the sand comes from the mass–balance equation

$$\rho_{Sand} = \phi\rho_f + (1 - \phi)\rho_s, \tag{11.18}$$

where ρ_f and ρ_s are the densities of the fluid and mineral, respectively.

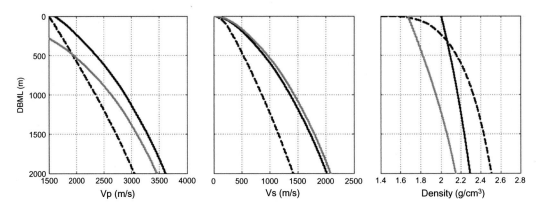

Figure 11.10 Depth trends for velocity and density. Solid black curves are for wet sand, dashed black curves are for mudrock, and gray curves are for gas sand.

It is important to emphasize that the dataset shown in Figures 11.8 and 11.9 is for well-sorted sandstone. Hence, it is the diagenetic or "cementing" trend. Based on this trend, we can model the elastic properties of poorly sorted sandstone using, for example, the constant-cement model.

11.10 Using trends at prospect location

After the recognition of key amplitude anomalies and associated DHIs in seismic data, it is important to link the plausible variations in rock properties to seismic responses. The inherent non-uniqueness of interpreting seismic data for rock properties can be mitigated if geological reasoning is used to reduce the number of variants in earth models. This can be done by perturbing the fundamental rock properties, such as porosity, mineralogy, and fluid; calculating the resulting elastic properties; and, finally, using these elastic properties in synthetic seismic generation. This geology-based approach helps constrain the range of the earth models by selecting porosity and lithology in a relatively narrow domain relevant to local geology (Dvorkin and Gutierrez, 2001). We then model the seismic response for multiple geological scenarios and generate a catalogue of seismic responses of potential exploration success and failure scenarios.

Following the model formulation for each rock type and fluid property, calibrated *P*-wave velocity, *S*-wave velocity, and density depth trends are generated for brine and hydrocarbon-bearing reservoirs scenarios using fluid substitution (Figure 11.10).

These depth curves are then used to generate half-space synthetic seismic responses for three sand bodies located at different depths in the OA under examination (Figure 11.11). At a certain depth, the normal-incidence responses from wet sand

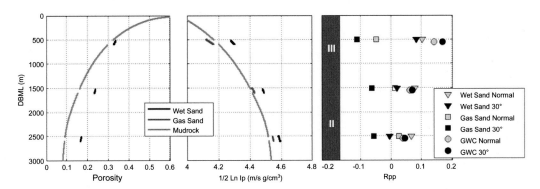

Figure 11.11 Left: Porosity versus depth for mudrock (gray) and clean sand (black). Middle: The respective impedance expressed as half the natural logarithm of impedance (see Eq. (4.1)). The mudrock curve is gray and sandstone curves are black for wet and dark-gray for gas cases (the latter is located to the left of the former). Right: Computed reflection amplitude for shale/sand interface and gas–water contact (GWC) at three different depths and for normal reflection and 30° angle of incidence. The separation of these responses decreases with depth. This example is only for the OA under examination. Vertical gray bar lists AVO classes for reflections at gas sands. The transition from Class III to Class II occurs at approximately 800m.

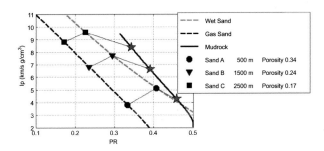

Figure 11.12 Impedance versus porosity cross-plot with the mudrock, wet sand, and gas sand curves. The symbols show the impedance and Poisson's ratio values at three depths and for the respective mudrock, wet, and gas sand bodies. Sands A, B, and C are the sands shown in Figure 11.11. Stars are for the elastic properties of the mudrock selected for Sands A, B, and C, respectively.

and gas sand become so close to each other that fluid recognition from seismic data becomes difficult if not impossible. This depth is called the DHI floor. In many basins the DHI floor is reached where the porosity of the sand becomes 0.15. In this prospect it is at a depth of about 1700 m (which can only be recognized, of course, given the high quality of the seismic data).

If the nearest well control is of reasonable quality and seismic impedance inversion is feasible, rock physics templates (Avseth *et al.*, 2005) can be used. Figure 11.12 shows such a template in the impedance versus Poisson's ratio plot using three basic rock types, the mudrock, wet sand, and gas sand.

Let us add in conclusion that in contrast to reservoir sands, the elastic properties of carbonate rock show little correlation with depth or age. They are strongly driven by diagenetic processes. Porosity and pore types are the main factors that control velocity, and variation in pore type is the main reason for variable velocity at fixed porosity. Initial lithology plus diagenetic alterations together control porosity and velocity evolutions from the time of deposition throughout the burial stages. Even pure-mineralogy carbonates exhibit wide ranges of velocity variation (Eberli *et al.*, 2003; Vanorio *et al.*, 2008; Fabricius *et al.*, 2010).

12 DHI validation and prospect risking

12.1 Introduction

A decision to drill an amplitude-based prospect must be based on rigorous validation of the seismic anomaly and evaluation of the prospect risk and the potential hydrocarbon volumes. Sometimes there is strong seismic evidence that hydrocarbons are present. Even in this case, careful DHI interpretation and risking of amplitude-supported prospects can reduce the exploration uncertainty and risk and, hence, encourage drilling a wildcat well.

Direct hydrocarbon indicator (DHI) interpretation consists of two phases: (a) recognition of seismic anomalies and (b) validation of the selected anomalies. The validation requires quantitative calibration of rock properties to seismic data (Chapter 11). Prospects with seismic data that contain clear apparent anomalies, possibly indicating a hydrocarbon presence, still have to undergo careful rock-physics-based examination to ensure that these anomalies are not false indicators.

Prospects without obvious amplitude anomalies are also studied using the same workflow to estimate the seismic response from prospective hydrocarbon-bearing reservoirs. If such investigation suggests that DHIs are expected, this exploration opportunity is highly risked and seismic acquisition and processing have to be revisited. On the other hand, if the study suggests that no obvious hydrocarbon indicators are expected, the possibility of hydrocarbon-bearing rocks in this particular prospect cannot be excluded. Such methodologies have been successfully applied to quantifying the exploration risk in many amplitude-supported prospects (Roden et al., 2005; Fahmy, 2006; Forrest et al., 2010; De Jager, 2012; Roden et al., 2012).

12.2 Feasibility studies

Quantitative interpretation (QI) techniques have been applied with frequent success to predicting lithology and fluids (e.g., Avseth et al., 2005). One of the key challenges facing geophysicists is to select the most appropriate seismic technology from the range of sophisticated options available, in order to attain maximum business impact. The level of sophistication of quantitative seismic interpretation and the complexity of the approach (e.g., seismic impedance inversion) are highly correlated with the effort and cost. It is recommended, therefore, that prior to embarking on a time-consuming and costly QI project, a feasibility study be carried out to fully understand the likelihood of attaining the desired outcome of the project. A checklist is supplied in the Appendix at the end of this book that can be used to scope a feasibility study and gather technical evidence of hydrocarbon presence.

Generally, the three main aspects a feasibility study must address are:
(a) General assessment of quality of the seismic data;
(b) Rock physics analysis of the reservoir and bounding rock properties; and
(c) Well-to-seismic ties and wavelet derivation.

12.3 Recognition of seismic anomalies

Seismic datasets are initially examined for any anomalies. Seismic anomalies are defined as abrupt changes in seismic character that can indicate variations in pore-fluid type (presence of hydrocarbons), lithology, and/or porosity. Some apparent anomalies can result from inappropriate seismic processing and they are considered seismic artifacts (Yilmaz, 2001). Seismic volumes can be screened in 2D by using maps and vertical profiles. In light of the availability of 3D seismic volumes, this screening technique has necessarily to be supplemented (or replaced) by effective reconnaissance of multiple and large volumes for potential hydrocarbons and lithology indicators. Using volume visualization and interpretation methods at every stage of a project's life cycle helps optimize the interpretation time, improve interpretation consistency and quality, and provide clear insights into the spatial structure of the subsurface.

Direct hydrocarbon indicators include bright spots, flat spots, dim spots, and character and phase change at prospective fluid contacts (e.g., oil/water or gas/water), as well as amplitude variation with offset (Ostrander, 1984; Shuey, 1985; Hilterman, 2001). Other DHIs may include polarity flips, amplitude shadows, low-frequency shadows, velocity anomalies involving a time delay (e.g., velocity sags), and gas chimneys (Brown, 2011). A seismic anomaly is not necessarily triggered by a commercial hydrocarbon accumulation. Amplitude anomalies can also be caused by low-saturation gas sands, clean blocky wet sands, and low-velocity mudrock. Confident identification of a hydrocarbon indicator necessarily involves accumulating all kinds of evidence, especially geological, at every phase of the interpretation.

To determine which seismic characteristics are relevant for specific geological settings, it is necessary to classify hydrocarbon-generated amplitude anomalies based on their geological environment. Most settings in clastic environments are accounted for using a classification based on four classes of amplitude variation with offset (AVO) responses from the top of a gas sand (Rutherford and Williams, 1989; Castagna et al., 1998). Class I or "dim spot" setting: very consolidated sands (porosity below 0.15) that have higher impedance than the bounding shale with the density contrast having a strong effect on the AVO response. Class II or "phase change" setting: moderately compacted sands (porosity between 0.15 and 0.25) where the impedance of gas sand and the bounding shale are approximately equal. Class III or "bright spot" setting: unconsolidated sand (porosity exceeding 0.25) that is encased in higher impedance shale. Class IV: unconsolidated sand similar to that in Class III, but bounded by high-velocity hard shale, siltstone, or carbonate.

Boundaries between AVO classes are gradational and strongly depend on the reservoir porosity and type of bounding lithologies. During the interpretation process, it is critical to associate an AVO class description with a particular rock type and pore fluid. For example, a Class II anomaly, in theory, can be related to the presence of hydrocarbons and can be interpreted as a low-porosity gas sand or a high-porosity oil sand. Hence, the term "Class II oil clean sand" is clearer than the ambiguous "Class II sand" identifier to describe an oil-bearing clean sand whose acoustic impedance is approximately equal to that of the bounding rock.

Regional screening of AVO anomalies requires the use of stacked AVO products, such as the envelope difference product (EDP). This product, also called the enhanced restricted gradient, is a robust measure of the change of amplitude with offset and it is very effective for scanning Class III and II anomalies. The envelope difference product is computed from the difference between the amplitude of the envelope of the near offset stack trace and that of the far offset stack trace, multiplied by the amplitude of the envelope of the far stack trace. The amplitude of the envelope is a seismic attribute also called the reflection strength (Taner *et al.*, 1979). Effective reconnaissance for potential fluid contacts often requires using flat spot enhancement techniques that improve the seismic feature visibility by adding parallel in-lines or cross-lines of 3D seismic data (called optical stacking). Evidence for fluid contacts includes changes in waveform character, amplitude switch-offs, variations in event dip, and structure-consistent amplitude anomalies due to tuning at the wedge created by an inclined reservoir top and the flat hydrocarbon–brine contact. It is important to identify and assess possible artifacts that could induce false flat spots (e.g., irregular illumination, absorption, multiples, and diagenetic and stratigraphic effects).

Spectral decomposition techniques allow us to extract the frequency spectrum and amplitude from seismic data and can also be used to fine-tune the identification of hydrocarbons when these techniques are properly integrated with AVO products in DHI analysis (e.g., Chen *et al.*, 2008). Spectral AVO attributes have been successfully used to extend the limits of hydrocarbon detection (Fahmy *et al.*, 2008).

12.4   DHI validation and prospect risking

A decision to drill an exploration prospect must be based on rigorous evaluation of the prospect risk and potential hydrocarbon volumes (Rose, 2001; De Jager, 2012). What is the chance that a well finds hydrocarbons and if so, how much? A geological assessment seeks to answer the following questions regarding the key prospect elements and estimate their probabilities:

- Is a geological trap indeed present?
- Is there a reservoir?
- Is there an effective seal for the hydrocarbons present?
- Is there a mature source rock that has generated these hydrocarbons and charged the reservoir?

In exploration, the probability of success (POS) or the probability that a well finds hydrocarbons is defined as the product of the individual probabilities of the four factors listed:

$$POS = P_{Trap} \times P_{Reservoir} \times P_{Seal} \times P_{Charge}. \tag{12.1}$$

The initial regional screening for amplitude anomalies is followed by an evaluation of the identified hydrocarbon and lithology indicators. This process requires a systematic, objective, and consistent effort to narrow the probability range of exploration success and the range of resource estimates. The interpretation workflow proposed by Roden *et al.* (2005) and Forrest *et al.* (2010) can assist in assessing the likelihood of hydrocarbon presence. This evaluation technique based on the scoring (grading) of the DHI's quality and robustness is applied during the recognition stage as well as during the validation of the anomalies that follows the recognition stage. The critical steps are to identify the AVO anomaly class, understand the geologic risk factors, review the seismic and rock physics data quality, and thoroughly analyze numerous seismic anomaly characteristics. Each potential amplitude-supported exploration opportunity is analyzed based on seismic data using the following general approach.

Prospect identification. This stage uses the key elements of the regional and local petroleum geology, such as trap, reservoir rock, seal, and hydrocarbon charge. Such knowledge should be assembled prior to amplitude "hunting" and validation.

It is necessary to document the prospect's location, type of hydrocarbon trap, terrain, depth to target, water depth, and expected reservoir extent and thickness. Data from geologically analogous fields and discoveries are valuable supplements to the site-specific knowledge. Yet, to avoid a costly mistake, analogues should not be relied upon without bringing them into a site-specific context. The interpretation of the seismic data addresses the trap description by supplying valuable images of the primary geological lineaments of the external reservoir geometry and their spatial associations. However, unless a DHI accompanies a potential hydrocarbon trap, the opportunity is usually rejected. Where a DHI is present, integrated mapping (structure plus amplitude anomaly) is conducted.

DHI validation. A checklist, supplied in Appendix A, can help gather evidence to support a DHI and its relation to the prospect. If the outcome of the validation is positive, a quick-look deterministic hydrocarbons volumetric estimation is conducted. Otherwise, the opportunity is rejected. Volume (V) calculation requires realistic values for seven parameters listed on the right-hand side of the following equation:

$$V = A \times T \times \tfrac{N}{G} \times \phi \times S_h \times \text{FVF} \times \text{RF}, \tag{12.2}$$

where A is the area of the hydrocarbon trap; T is the reservoir thickness; $\tfrac{N}{G}$ is the net-to-gross ratio; ϕ is the porosity; FVF is the formation volume factor (the ratio of the hydrocarbon volume at atmospheric conditions to that at the reservoir conditions); and RF is the recovery factor (Rose, 2001).

The scoring approach (Roden *et al.*, 2005; Forrest *et al.*, 2010) is applied during this recognition stage and the posterior validation of the anomalies. This is essentially a qualitative assessment during which the structure-related amplitude changes are identified, classified, and evaluated. The key characteristics that define the quality of a potential DHI include:

- Down-dip conformance on a depth structure map (possible gas/oil/water contact).
- Consistency (versus patchiness) of the amplitude in the target area (on full stack or far-offset data).
- Quantifying the seismic response of wet sand versus sand with hydrocarbons.
- Signature match (polarity and confidence in the phase of the seismic data).
- Proven DHIs at nearby locations (analogue successes in the same general area).
- AVO studies (observations, comparison between those at the prospect and off-structure).
- Flat spots (possible hydrocarbon–water contacts).
- Phase/amplitude change at the edge of the anomaly (possible fluid contacts).

It is also crucial to quantify common DHI pitfalls (false positives), including (a) soft shale, ash, coal, or marl; (b) high-porosity clean wet sand; (c) wet sand with low residual gas saturation; (d) hard streak above or between two wet sandstones; (e) amplitude caused by sharp onset of overpressure; (f) presence of salt and/or lithology switches to, for example, carbonate (also polarity or phase issue); (g) presence of CO_2 or nitrogen; (h) tuning effects; and (i) lateral lithology or thickness changes and associated imaging problems.

An amplitude anomaly means relatively high or relatively low amplitude where both can be associated with hydrocarbons. For instance, typical Class III anomalies show on-structure brightening, while Class II and I anomalies usually show on-structure dimming on near-offset data.

Figures 12.1 and 12.2 show a comparison between volume AVO observations with modeled seismic responses verified by the examination and analysis of migrated gathers. In this example, the identified Class III AVO anomaly is located at about 500 m DBML and corresponds to the modeled top clean gas sand scenario (Figure 11.12). Depending on the local geology, relatively shallow gas accumulations can be considered

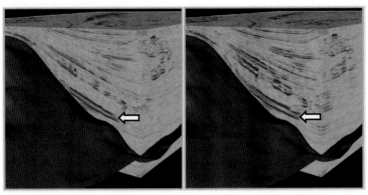

Sail - Near Stack Volume   Sail - Far Stack Volume

Figure 12.1 AVO volumes for near and far stacks. SAIL (seismic approximate impedance log) stack is obtained by integrating the appropriate seismic trace with respect to the travel time (Waters, 1992). Red is for negative while blue is for positive amplitude. For color version, see plates section.

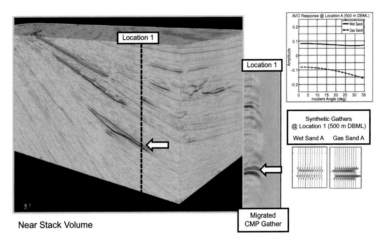

Figure 12.2 Comparison of near-stack volume, migrated gather, and synthetic seismic responses for a wet and gas clean sands scenarios at 500 m DBML (Gutierrez and Dvorkin, 2010). For color version, see plates section.

a potential economic target or a hazard for drilling operations (Dutta *et al.*, 2010). Amplitude analysis is conducted in a similar way for both prospect evaluation and shallow hazards assessments.

Amplitude calibration. At this point we have established that hydrocarbons are possibly present in the geological trap. However, it is still unclear what kind of hydrocarbon (oil or gas) is associated with the DHI. The use of seismic amplitude and rock physics depth trends (Chapter 11) is a quantitative tool for predicting the pore-fluid type in an exploration area where the nearest well control is some distance away. Calibration between the real seismic and synthetic amplitude cannot be conducted directly because the seismic data has an arbitrary scaling factor. This is why we have to deal with the amplitude ratio. Specifically, we need to ensure that

$$\frac{A}{B}\bigg|_{Seismic} \approx \frac{A}{B}\bigg|_{Synthetic} , \qquad (12.3)$$

where A is the amplitude at the seismic anomaly and B is that of the background. This technique is called A/B normalization. B is usually estimated over a large time window computing an RMS amplitude over numerous seismic traces. Sometimes B is defined by estimating the average amplitude for the potential brine-saturated reservoir located down-dip of the anomalous amplitude location (Hilterman, 2001). This method can also be used with band-limited inversion products (Connolly, 1999; Lancaster and Whitcombe, 2000).

If the nearest well control is reliable and seismic inversion is feasible, the use of rock physics templates (Figure 11.12) to interpret relations among elastic attributes (e.g., acoustic and elastic impedance) in terms of rock properties, as well probabilistic calibration techniques can be helpful (e.g., Avseth *et al.*, 2005).

Probabilistic amplitude analysis of the stacked amplitude in terms of pore fluid is divided into three steps:

(a) Stochastic forward modeling (Monte Carlo based predictions) of different effects (lithology, porosity, and pore pressure and fluid) on the seismic response. Quantitative seismic parameters are connected to rock and fluid properties through rock physics models.

(b) Calibration of the modeled data with the measured amplitudes in the zone of interest to compute an appropriate scaling factor.

(c) Mapping of the probability of different pore fluid scenarios. If only liquid hydrocarbons are of commercial interest and the likelihood of oil is high, static and dynamic subsurface models are generated. Otherwise (e.g., high probability of gas presence) the opportunity is rejected.

Commercial hydrocarbon volume threshold. Minimum commercially viable compartment size is usually estimated locally using commercial factors. If the expected volume is less than this threshold, the opportunity is rejected. Finally, if the results of the reservoir dynamic modeling are encouraging, a well drilling proposal is formulated and presented.

Part V

Advanced rock physics: diagenetic trends, self-similarity, permeability, Poisson's ratio in gas sand, seismic wave attenuation, gas hydrates

13 Rock physics case studies

In this chapter we present findings from laboratory and well data analysis showing potentially useful extensions of rock physics diagnostics and data analysis.

13.1 Universality of diagenetic trends[*]

Honoring stratigraphic constraints guarantees that a rock physics trend is deposition- and site-specific. *Rationalization* by effective-medium modeling makes a trend general, determines the domains of its applicability, and thus reduces the risk of using the trend outside the initial data range. As an example of rationalization, consider work by Avseth *et al.* (2000) in which empirical data trends are supported by effective-medium curves that represent varying sorting and cementation. Such melding of data and theory is a signature of modern rock physics, where the goal is not only to observe and relate but also to explain and generalize. How general are rational rock physics trends? Can they be applied across tiers of deposition and across geographic areas? These are the questions addressed below (see the original publication by Dvorkin *et al.*, 2002).

Consider first a vertical North Sea well whose log curves of gamma-ray (GR), porosity, and the compressional (M) and shear (G) moduli are shown in Figure 13.1. These elastic properties are recalculated from the in situ data by fluid substitution.

The pay zone is the flat-GR interval at about 1.8 km depth. Both elastic moduli in this pay zone are much larger than those in the surrounding rock. The observed relatively high stiffness of the pay zone is due to the small amounts of amorphous quartz that cements the grains at their contacts (Avseth, 2000). Such cementation acts to strongly increase the velocity while keeping the porosity high. The strong elastic contrast between sand and shale translates into the two distinctive modulus–porosity trends shown in Figure 13.2. The lower branches in those plots are called the uncemented or friable trends and the upper branches are the contact cement trends.

---

[*] This part was modified from work originally published by SEG (Dvorkin *et al.*, 2002)

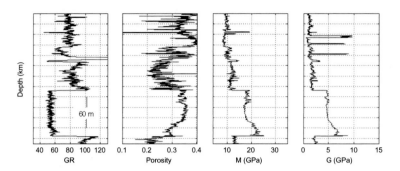

Figure 13.1 Well log curves for the North Sea well. The elastic moduli are recalculated for full water saturation from the original data.

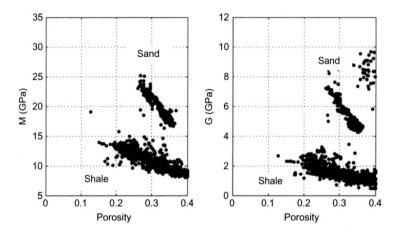

Figure 13.2 Cross-plots for the data shown in Figure 13.1. Compressional (left) and shear (right) modulus versus porosity. The outliers apparent in the upper right part of the shear-modulus graph are not taken into consideration in the following analysis.

Consider next a well located on the US Gulf Coast, several thousand miles from the North Sea well (well log curves are shown in Figure 13.3). As in the North Sea well, the elastic moduli are theoretically recalculated for 100% water saturation. The pay zone is the low-GR interval at a depth of about 3.5 km, some 1.7 km deeper than the North Sea interval. The elastic modulus versus porosity cross-plots for the Gulf Coast well (Figure 13.4), similar to those for the North Sea well, exhibit clear separation between the pay zone and the shaly remainder of the well.

Consider now a superposition of cross-plots for the two wells (Figure 13.5, top). In spite of the large geographic and depth difference between the wells, the elastic moduli versus porosity trends for the Gulf Coast well almost seamlessly continue the trends for the North Sea well. The trends are approximately supported by the constant-cement model for the pay zones and the soft-sand/shale for the shaly intervals (Figure 13.5,

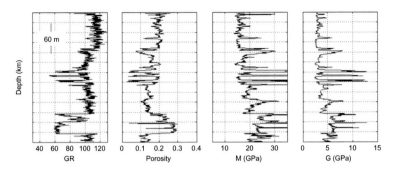

Figure 13.3 Same as Figure 13.1 but for the Gulf Coast well.

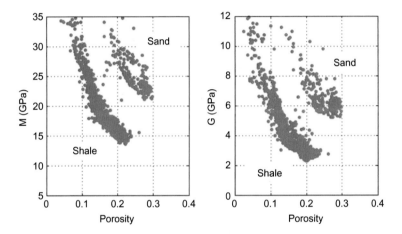

Figure 13.4 Same as Figure 13.2 but for the Gulf Coast well.

bottom). The burial-related compaction and cementation processes in the Gulf Coast well continue those started in the North Sea well.

Let us next examine Han's (1986) dataset shown in Figure 1.8. These data include more than 60 consolidated sandstone samples from the continental United States and Gulf Coast. The velocity is plotted versus porosity in Figure 13.6 (left). The data displayed have been collected on room-dry samples at 40 MPa differential pressure. The velocity displayed is theoretically calculated for full water saturation conditions from the room-dry data via Gassmann's fluid substitution. The cross-plot in Figure 13.6 (left) cannot be directly used for relating velocity to porosity because it includes several separate velocity–porosity trends. These trends, defined by the amount of clay in the rock, are shown in Figure 13.6 (right). Once the clay content is constrained, clear velocity–porosity trends appear.

Figures 13.7 and 13.8 display well log data from two wells, AC5 and AC6, located in the Acae Field, Colombia. The producing zone in both wells is the Caballos Formation located below 3.2 km depth. The higher-GR zone above is the Villeta

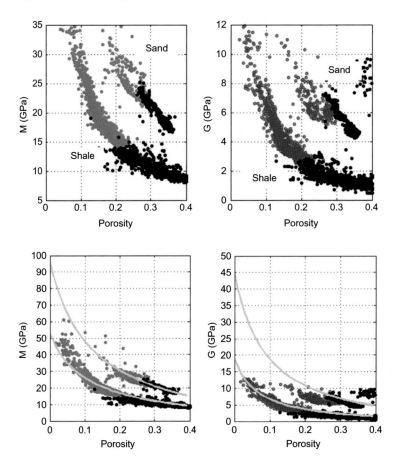

Figure 13.5 Top: Figures 13.2 and 13.4 superimposed. The North Sea data points are colored black while the Gulf Coast data points are colored gray. Bottom: The same cross-plots but in a wider elastic modulus scale. Gray curves are from the constant-cement model (sand) and soft-sand/shale model (shale). For the constant-cement model we used the soft-sand functional form with the coordination number 15; critical porosity 0.40; shear modulus correction factor 1; and differential pressure 30 MPa. The bulk modulus and density of water were 3.0 GPa and 1.0 g/cm³, respectively. The mineralogy was pure quartz. The inputs for the soft-sand model were the same except for the coordination number 6 and mineralogy, which was in this case 50% quartz and 50% clay. We did not change these inputs between the North Sea and Gulf Coast wells in order to obtain continuous model curves.

Formation, marine shale. The velocity–porosity cross-plots for AC5 and AC6 for the acceptable-data-quality intervals selected based on the caliper log are shown in the plots next to the depth plots. The data form two definite trends. The upper trend is for the low-clay-content Caballos sandstone while the lower trend is for the Villeta shale. The superposition of Han's data on the Acae cross-plots indicates that the Caballos

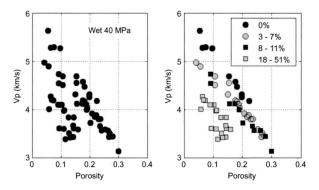

Figure 13.6 Velocity–porosity cross-plots for Han's data at 40 MPa differential pressure. The values shown are calculated for water-saturated rock from room-dry data. Left: Data in a wide range of clay content variation. Right: trends for narrow clay content ranges, as shown in the legend.

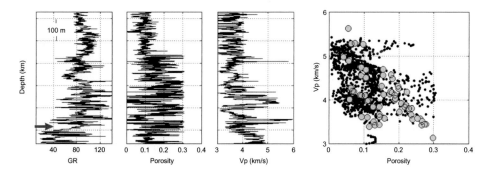

Figure 13.7 First three tracks are depth plots for GR, porosity, and the *P*-wave velocity in Acae well AC5. The fourth plot is a velocity–porosity plot for reasonable-quality AC5 data (black symbols) with Han's data superimposed (gray circles, same for the entire clay content range). The arrow in the GR track points at the top of the Caballos formation.

data mimic the low-clay-content Han trends and the Villeta data fall on top of the high clay-content Han data. In spite of the geographic separation, the rock physics behavior of a Colombian sand/shale system is similar to that of the North American rocks.

By analyzing rock physics trends from wells located in different parts of the world, one may find a striking similarity among them. This similarity confirms the fact that there are only a few natural forces, gravity, water and air transport, temperature, and pressure that shape rock, and that these natural forces are quite similar among different basins. The practical lesson is that data from geographical locations not necessarily close to the prospect may be used as analogues and also for quality control of newly acquired data.

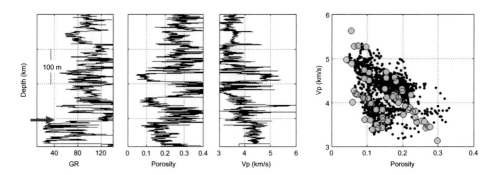

Figure 13.8 Same as Figure 13.7 but for well AC6.

13.2   Self-similarity in rock physics[*]

Self-similarity in mechanics means that a function of more than one argument is fully determined by a combination of these arguments. Gal *et al.* (1999) presented an example of such behavior in consolidated sandstone where both I_p and I_s (as well as V_p and V_s) uniquely depend on a linear combination of the total porosity and clay content, $\phi + aC$ (the porosity of the load-bearing frame). Such dependence simply means that it is impossible to jointly resolve I_p and I_s data for ϕ and C. Instead only the $\phi + aC$ combination can be found. This effect is illustrated in Figures 13.9 and 13.10 where Han's (1986) velocity–porosity trends, which have scatter due to the varying C, collapse to tight trends if V_p and V_s are plotted versus $\phi + 0.3C$. This may mean that the clay in the pore space of these samples is not load-bearing, so that it occupies the space but contributes very little to the stiffness of the rock.

A well-known example of self-similarity in fluid dynamics is viscous flow, where the characteristics are determined by the Reynolds number, Re $= \rho VL / \mu$, where ρ and μ are the density and dynamic viscosity of the fluid, respectively; V is the velocity of the fluid flow; and L is the linear scale of the flow (such as the diameter of the conduit). Self-similarity (or the dimensional analysis) is commonly used in mechanics to evaluate a problem without actually solving it. It has not been given any significant consideration in rock physics, except for the recognition of the fact that the absolute permeability depends on the combination of porosity and grain size, d: ϕ / d^2 (Mavko *et al.*, 2009), simply because the number of variables measured in the field is usually very small.

The examples given in Figure 13.10 present self-similarity as an obstacle in resolving field data for reservoir properties: we can only interpret the impedance in terms of a linear combination of the porosity and clay content but cannot resolve for these

---

[*] This part was modified from work originally published by SEG (Dvorkin, 2007)

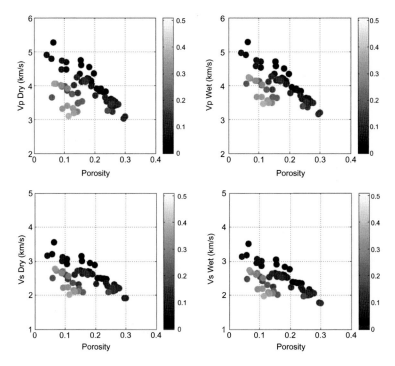

Figure 13.9 Han's (1986) data at 40 MPa confining pressure for dry (left) and wet (right) conditions. The *P*-wave velocity is plotted in the top row while the *S*-wave velocity is in the bottom row. The color represents the clay content.

two variables separately. However, it can be also viewed as an advantage. Indeed, it can help establish the requirements for data accuracy (e.g., no matter how accurate the seismic I_s is, it still may not help in reservoir quality discrimination). Self-similarity, once established, can also help design a field experiment that may solve the problem at hand, or at least clarify which links among rock properties must be established and which assumptions must be made to solve the problem.

Two-dimensional velocity–porosity plots, color-coded by *C*, are in essence 3D plots, where the third dimension is the color. Figure 13.11 shows $I_p = \rho_b V_p$ and $I_s = \rho_b V_s$ versus ϕ and *C* plots of the data from Figure 13.9 at two different view angles. It is not surprising, that by rotating this 3D plot we can find a projection that eliminates the scatter due to clay content variability and thus puts all the data points onto a tight trend. What is remarkable is that the same rotation *jointly* puts both I_p and I_s onto tight trends. This is an illustration of self-similarity where the *P*- and *S*-wave properties of the rock depend on the same linear combination of the total porosity and clay content.

Rock physics transforms also exhibit self-similarity. Consider Figure 13.12, where the Raymer–Dvorkin (Chapter 2) curves for the wet rock are superimposed upon the data shown in Figure 13.9. In Figure 13.13 we plot these model curves versus $\phi + 0.3C$.

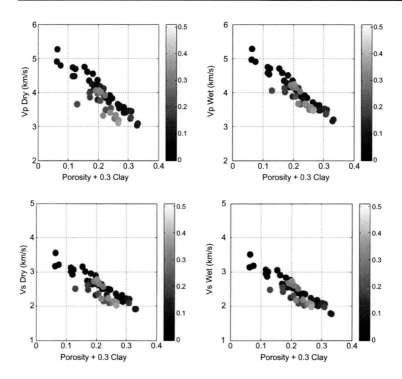

Figure 13.10 Same as Figure 13.9 but for the velocity plotted versus a linear combination of porosity and clay content.

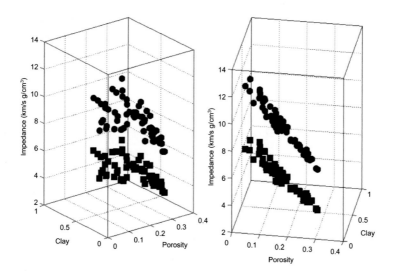

Figure 13.11 Data from Figure 13.9 in 3D. Rotation helps put the data onto tight trends (right).

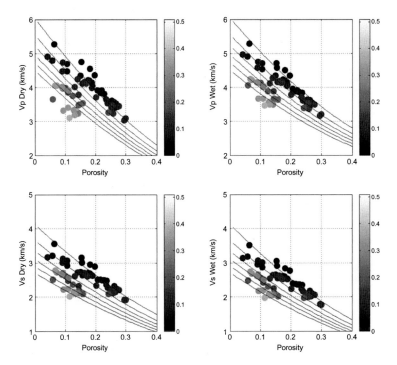

Figure 13.12 Data from Figure 13.9 with the Raymer–Dvorkin model curves superimposed. Each curve is for a fixed clay content, starting with zero for the top curve and ending with 50% for the bottom curve.

As a result, these velocity–porosity–clay transform curves simultaneously collapse into tight trends for both P- and S-wave velocity.

The same self-similarity behavior can often be found in well data. Consider log data from a well that penetrates several wet and gas-saturated sand intervals as well as shale (Figure 13.14). Both the gas and wet sand intervals form tight and separate impedance–porosity trends in the impedance–porosity cross-plots (Figure 13.15). After calculating the elastic properties in the entire interval for fully water-saturated conditions using Gassmann's fluid substitution, we find that (now with the same fluid in the pore space) both gas and wet sand form a practically single tight trend in the impedance–porosity panel (Figure 13.15).

Figure 13.16 shows that this trend can be mimicked by the soft-sand model. The model curves in Figure 13.16 (left) are plotted for wet conditions. Each curve is for a fixed clay content starting with zero (the upper curve) and ending with pure clay (the lower curve), with intermediate curves drawn at clay-content increment 0.2. As in the previous example with Han's laboratory data, these well data collapse to a single tight impedance–porosity trend if the wet-rock impedance is plotted versus $\phi + 0.15C$, where the clay content is calculated by linearly scaling the gamma-ray curve. So do the soft-sand model curves.

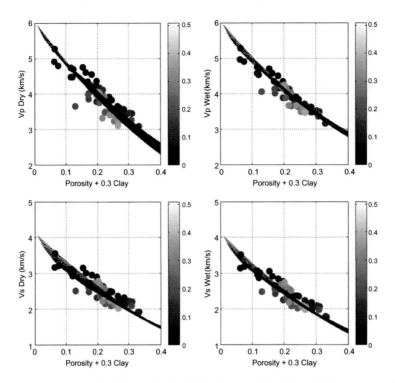

Figure 13.13 Data from Figure 13.9 with the Raymer–Dvorkin model curves superimposed. Each curve is for a fixed clay content starting with zero for the top curve and ending with 50% for the bottom curve.

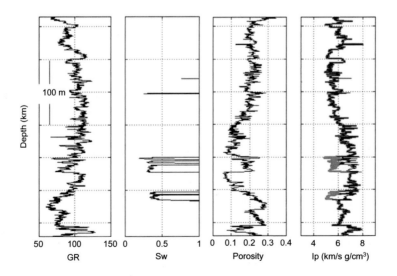

Figure 13.14 Well log curves from a gas well. From left to right: Gamma-ray; water saturation; porosity, and the *P*-wave impedance. The black curve in the impedance frame is calculated for wet-rock conditions while the bold gray curve is as recorded in situ.

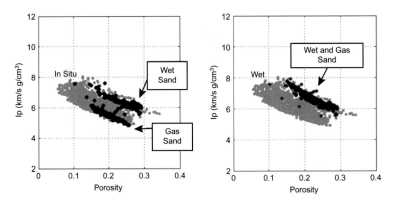

Figure 13.15 Impedance–porosity cross-plots using the data from Figure 13.14. Left: In situ conditions. Right: Wet conditions. The sand (gas and wet) is highlighted black.

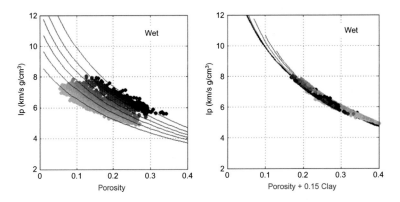

Figure 13.16 Wet-rock impedance data from Figure 13.15. Left: Impedance color-coded by clay content versus porosity. Right: The same data versus porosity plus 0.15 times clay content. The curves are from the soft-sand model as explained in the text.

The main lesson from this study is that the availability of *P*- and *S*-wave data does not necessarily translate into the ability to jointly resolve for two reservoir parameters, such as porosity and clay content. The reason is self-similarity, which is the dependence of both measurements on the same combination of porosity and clay content. This statement by no means implies that *S*-wave data are redundant and not needed in seismic characterization. Indeed, the relative or absolute seismic attributes, such as Poisson's ratio, the V_p / V_s ratio, and the fluid factor, which can be derived from joint *P*- and *S*-impedance inversion, are invaluable in fluid and lithology detection and delineation. It is within a geobody thus delineated that self-similarity may be an obstacle to resolving two seismic attributes (I_p and I_s) for two bulk properties of the sediment (ϕ and *C*). A solution is to establish, understand, and utilize geology- and deposition-imposed links between the reservoir parameters in question, that is, venture

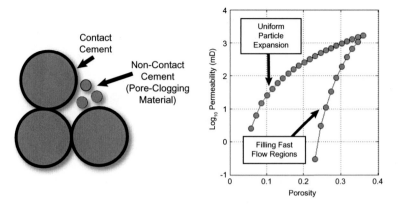

Figure 13.17 Left: Large grains with contact-cement rims around them and small (non-contact-cement) particles clogging the pore space. Right: Permeability–porosity trends computed on a Finney pack using two porosity reduction scenarios: (a) uniformly expanding each grain, which amounted to growing contact cement around them and (b) placing small particles inside the pores away from large grains.

beyond the seismic attribute calculus into the realm of sedimentology where missing equations can be found that link porosity, clay content, saturation, and even permeability (see next section).

13.3 Elastic properties of rock and its permeability[*]

Dvorkin and Brevik (1999) examined well data combined with permeability measurements on selected samples from the logged interval. The permeability–porosity data do not form a trend. After conducting the rock physics diagnostics on the well data, and using the contact-cement and soft-sand models, it was possible to estimate the amount of non-contact cement (Figure 13.17, left) in each sample based on their elastic modulus versus porosity datum's location between the two model curves. Once the fraction of the non-contact cement was determined from the modulus–porosity data, the measured permeability was plotted versus this fraction and formed a fairly tight trend. The reason for the emergence of a trend is that the material contained inside the pores away from the large grains presents the strongest obstacle to viscous fluid flow: because the fluid has to stick to the pore-space walls (the no-slip boundary condition), it is strongly and negatively influenced by the increasing pore-surface area. The small particles inside the pore space act to strongly increase this area as opposed to the material that envelops the grains thus forming the contact cement.

---

[*] This part was modified from work originally published by SEG (Expanded Abstract, Grude *et al.*, 2013)

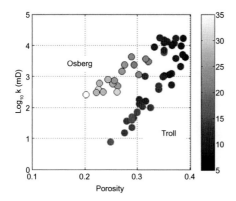

Figure 13.18 Permeability versus porosity for the Oseberg and Troll datasets color-coded by the dry-rock compressional modulus (GPa) shown in the color bar.

An example of permeability–porosity trends computed on a 3D model of rock formed by a dense random pack of identical grains with the additional material placed into the pore space is shown in Figure 13.17, right. This result was provided by Bosl *et al.*, 1998. An efficient computational engine used for this purpose was the lattice–Boltzmann method that mimics the Navier–Stokes equations for a Newtonian fluid flow with no-slip boundary conditions. In that work, the original building block used for simulation was the Finney pack, a random dense pack of identical spheres with porosity of about 0.36. The porosity was lowered from this initial point by two different procedures: (a) each spherical particle was uniformly expanded with the resulting sphere overlapping and (b) the small solid particles were placed where the flow velocity was the highest (also in a step-wise fashion). Because in the second scenario the fastest flow developed away from the large grains, this porosity reduction model was essentially putting small particles in the middle of the pore space, thus clogging the main conduits.

Coincidentally, the elastic moduli of the model rock with contact cement were much larger (at the same porosity) than those where the particles were placed inside the pore space (non-contact cement) – see also Chapter 2. This means that the elastic modulus can, in principle, be used to assess the permeability. An illustration of this principle is in Figure 13.18 where we plot permeability versus porosity for two North Sea datasets: the Oseberg field where the grains were contact-cemented and the Troll field where the small particles clogged the pore space (Strandenes, 1991; Blangy, 1992).

The same principle was used by Grude *et al.* (2013) in application to well data from a Barents Sea well penetrating a wet sandstone reservoir (Figure 13.19) where the porosity and permeability were measured on samples extracted from the core material.

The rock physics diagnostics conducted on the well data indicate that the constant-cement model curves with varying degree of the contact cement accurately describe the data: at the same porosity, the samples with higher contact-cement content are

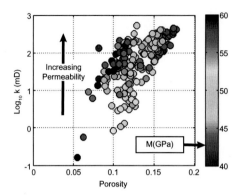

Figure 13.19 Permeability versus porosity color-coded by the compressional modulus. After Grude *et al.* (2013). For color version, see plates section.

stiffer and contain less fines in the pore space. This, in turn, affects the permeability: at the same porosity, stiffer samples have higher permeability (Figure 13.19).

A lesson from this case study is that model-based in-depth rock physics analysis of well log and laboratory data can yield "extra mileage" and help us categorize the permeability of rock under examination. Moreover, Dvorkin and Brevik (1999) indicate that the theoretically inferred amount of contact cement is correlated with the rock strength as the observations from that case study show that the lower the amount of the contact cement the more prone to sanding is the wellbore.

13.4 Stratigraphy-constrained rock physics modeling

Cycles in stratigraphy. Stratification is a common characteristic of sedimentary rocks. Cyclic and episodic processes and events as well as biologic and diagenetic overprint are responsible for stratification. Strata usually show rhythmicity (cyclicity), due to regularly alternating beds as well as repeated series of particular lithologies, which are called depositional cycles (Einsele *et al.*, 1991).

Genetically, depositional cycles are controlled by two groups of sedimentary mechanisms: autogenic (within-basin) and allogenic (extra-basinal). For instance, autogenic examples in fluvial systems are the migration and superposition of channel systems. In contrast, allogenic sequences are caused by changes external to the basin, such as climatic variations and tectonic movements (Miall, 1997).

Sedimentary processes, including cyclicity and the resulting depositional features, are examined at different geological temporal and spatial scales. Miall (1996) suggests a classification of reservoir architectural units and their associated sedimentary processes according to the depositional time scale into 10 classes spanning at least 12 orders of magnitude, from the lamina at one extreme to a basin-fill complex at the other

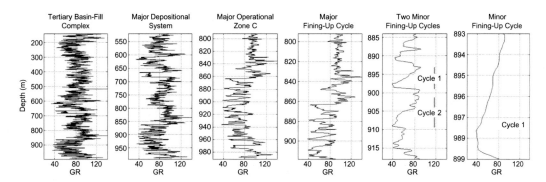

Figure 13.20 Vertical GR profiles in an LCI oil well with the scale of investigation reducing from left to right, as explained in the text.

extreme. A hierarchy of spatial scales spans at least 14 orders of magnitude, from a few centimeters for a ripple to tens of kilometers for a major sedimentary basin.

Elemental sedimentary cycles can be affected by several diagenetic processes, including mechanical and chemical processes during burial: mechanical compaction, cementation, replacement of unstable minerals, dolomitization, and dissolution (Blatt *et al.*, 1980; Boggs, 1995).

Siliciclastic reservoirs: a fluvial deposit example. Gutierrez (2001) examined well data from highly heterogeneous reservoirs in Tertiary fluvial sandstones in the mature giant La Cira-Infantas oil field (LCI) in Colombia. One of the key lessons was that the rock physics trends become tighter and more meaningful if subsets of log and core data examined are constrained by a sequence stratigraphy framework.

The siliciclastic reservoirs in the LCI belong to the Colorado (Zone A) and Mugrosa formations (Zones B and C). Oil production comes from loosely consolidated Tertiary sands. The reservoir rocks are fine- to medium-grained, subarkosic, clean-to-shaly sandstones. Average porosity is 20% in Zone B and 23% in Zone C. Permeability may reach 1500 mD. It is highly variable laterally and vertically (Dickey, 1992).

We concentrate here on the Mugrosa formation reservoirs. Zone B comprises a thick succession (550 m) of alternating fine-grain sandstones and shales. The lower part of the sequence consists of dull blue and brown, massive, mottled shales with some thin inter-beds of fine-grained sandstones and light-green sandy shales. The upper part of the formation is gray, fine- to coarse-grained, somewhat pebbly sandstone, inter-bedded with minor amounts of green and mottled shales (Morales *et al.*, 1958). Zone C is the most prolific reservoir in the LCI. This reservoir zone is gray to grayish green, fine- to medium-grained, rarely coarse and pebbly sandstones, with inter-bedded gray, blue, and red shales (Morales *et al.*, 1958). Sequence stratigraphy indicates that the depositional facies in the LCI Tertiary rocks are associated with fluvial channel systems. Zone B represents mixed load channels or meandering

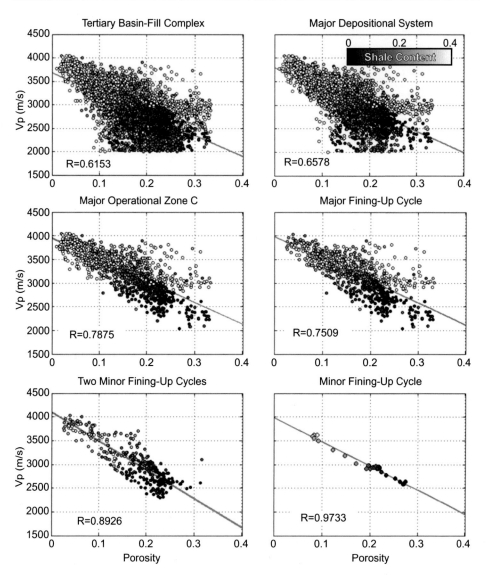

Figure 13.21 Velocity versus density-derived porosity for the six depositional intervals shown in Figure 13.20. The data points are color-coded by the shale content (derived from GR). The straight lines in each subplot are the best-linear-fit to the data displayed. The respective correlation coefficient is listed in each subplot.

systems, while Zone C was deposited as bed load channels or braided stream systems (Laverde, 1996).

Figure 13.20 shows vertical GR profiles in one of the wells at gradually reducing depth intervals, spanning a wide spatial scale spectrum, from a Tertiary basin-fill complex to its major depositional subsystem to the major operational zone (Zone C in LCI)

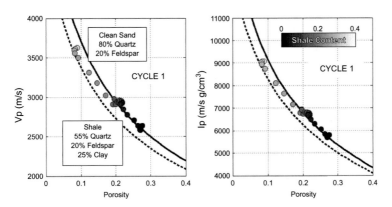

Figure 13.22 Velocity (left) and impedance (right) versus density-derived porosity for Cycle 1 from Figure 13.20. The model curves are from the soft-sand model as explained in the text. The mineralogy used is listed in the plot on the left.

to its major fining-upwards cycle to two separate fining-upwards cycles within and, finally, to one of these two cycles (Cycle 1).

Figure 13.21 shows velocity–porosity cross-plots for the six depth intervals displayed in Figure 13.20 in order of decreasing vertical extent of the interval under examination and the number of elemental cycles contained in this interval. The correlation coefficient increases as smaller and smaller cycles are selected within which the sedimentary environment conditions and diagenetic nature of the sands are more uniform. The final cross-plot in Figure 13.21 (bottom right) shows a high-correlation trend for a single well-developed fining-up cycle between 893.21 and 898.48 m deep.

Rock physics modeling. To match and explain the velocity–porosity behavior of the depositional cycles, the soft-sand model of Dvorkin and Nur (Mavko *et al.*, 2009) was used. This choice is based on the fact that the sand under examination appears to be uncemented and much softer than typical consolidated sandstone in the same porosity range. Figures 13.22 and 13.23 show velocity–porosity cross-plots for Cycle 1 from Figure 13.21.

In the velocity–porosity and impedance–porosity Cycle 1 plots (Figure 13.22), the upper curves are the soft-sand model for clean sand with the mineralogy 80% quartz and 20% feldspar. They are computed for 100% wet rock and 12 MPa differential pressure. The lower curves in the same plots are for shale with mineralogy 55% quartz, 20% feldspar, and 25% clay. It is now apparent that the velocity–porosity trend observed in these data is not a simple linear trend. The data points move from the clean-sand model curve to the shaly-sand curve as porosity decreases and shale content increases. The same cross-plots but for Cycle 2 are shown in Figure 13.23. The clean-sand model curve used in modeling Cycle 1 matches the clean-sand Cycle 2 data. However, to match the shale data in this cycle, we need to change the shale mineralogy and assume that now it is 20% quartz, 20% feldspar, and 60% clay. This seemingly

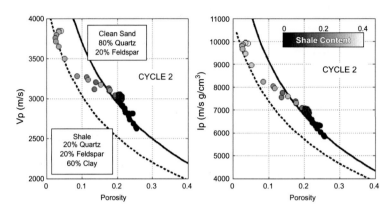

Figure 13.23 Same as Figure 13.22 but for Cycle 2.

arbitrary selection of mineralogy is consistent with the grain size and classification of deposits in the intervals selected. In Cycle 1 the lithology varies from clean sand to siltstone, while in Cycle 2 the lithology changes from clean sand to claystone.

Lessons. It is apparent from our modeling that the impedance of the high-shale-content rock is higher than that in the low-shale-content rock. This observation seemingly contradicts an established notion that increasing clay content acts to reduce the rock's velocity (see Figure 1.8). To explain the effect observed in the LCI data, we need to recall that (a) the porosity in this shale is smaller than that in the sand and (b) the clay content in this shale may be as low as 20% to 30%.

In fact, the shale in LCI actually contains a large fraction of silt that has mineralogy close to that of clean sand. The main difference between shale and sand in this case is the grain sorting, which deteriorates in the shaly layers. This poor sorting in shales acts to reduce the total porosity (compared to clean sands) and, as a result, acts to increase the velocity.

This investigation emphasizes the importance of understanding local depositional conditions and tailoring rock physics models accordingly. Although this conclusion may be perceived as contradictory to the section of this chapter discussing the universality of diagenetic trends, it is not, because the same but *locally adjusted* rock physics models can be effectively used at various geographic locations

14 Poisson's ratio and seismic reflections*

14.1 The high Poisson's ratio issue in gas sand

Dvorkin (2008c) discusses the disparity in Poisson's ratio (v) values measured in gas-filled sand in the well and laboratory. Laboratory data supported by granular-medium and inclusion theories indicate that the Poisson's ratio in gas-saturated sand lies within a zero to 0.25 range, with typical values about 0.15. However, some well log measurements, especially in slow gas formations, persistently produce a Poisson's ratio as large as 0.3. The questions addressed in this chapter are (a) whether high Poisson's ratio values measured in gas sands are realistic and have physics-based explanations and (b) how much these Poisson's ratio values affect the seismic response. Of course, the present discussion is based on the assumption that the well data are of high quality.

Poisson's ratio (v) can be calculated from any pair of independent elastic constants, such as bulk (K) and shear (G) moduli or the two Lamés constants (λ and $\mu \equiv G$, respectively). It can also be calculated from V_p and V_s, which are routinely measured in the laboratory and/or well. Operating with v instead of its qualitative equivalent, the velocity ratio (V_p / V_s), is convenient because the former is contained between zero and 0.5 while the latter may span the range between $\sqrt{2} \approx 1.142$ (for $v = 0$) and $+\infty$ (for $v = 0.5$).

Poisson's ratio (or rather its contrast) is important in seismic interpretation as it explicitly affects the AVO response of the reservoir (Chapter 4): Hilterman's (1989) approximation to the Zoeppritz reflectivity equations shows that the P-to-P reflectivity, R_{pp}, at an angle θ is proportional to the v contrast between the lower and upper elastic half-space:

$$R_{pp}(\theta) \approx R_{pp}(0)\cos^2\theta + 2.25(v_2 - v_1)\sin^2\theta, \tag{14.1}$$

where v_1 corresponds to the upper half-space while v_2 corresponds to the lower one.

To illustrate the effect of v on the AVO response, let us examine the reflection amplitude at the interface between hypothetical shale and gas-sand layers with the total

---

* This part was modified from work originally published by SEG (Dvorkin, 2008)

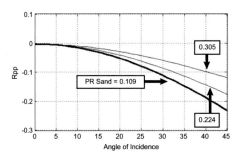

Figure 14.1 *P*-to-*P* reflectivity versus the angle of incidence at a shale/gas-sand interface (example in the text). The bold curve is for the original case where the Poisson's ratio in the sand is 0.109. The uppermost curve is for the extreme case where the Poisson's ratio in the sand is 0.305. The curve in the middle is for the case where the Poisson's ratio in the sand is 0.224.

porosity 30% in the shale and 25% in the sand. The clay content is 90% in the shale and 10% in the sand. The shale is filled with brine whose bulk modulus and density are 2.75 GPa and 1.02 g/cm³, respectively. The sand has 30% water saturation, with the rest of the pore space filled with gas whose bulk modulus and density are 0.07 GPa and 0.21 g/cm³, respectively.

The soft-sand model (Chapter 2) gives the following values for V_p, V_s, and the bulk density: 2.14 km/s, 0.83 km/s, and 2.12 g/cm³, respectively, in the shale and 2.15 km/s, 1.42 km/s, and 2.10 g/cm³, respectively, in the sand. Poisson's ratio is 0.412 in the shale and 0.109 in the sand. The *P*-to-*P* reflection amplitude versus the angle of incidence computed according to the exact Zoeppritz (1919) equations is shown in Figure 14.1.

Assume next that the V_s measured in the sand is erroneous and is only 90% of that computed using the soft-sand model, that is, 1.28 km/s instead of 1.42 km/s. The corresponding v is now 0.224 instead of 0.109. The respective AVO curve (Figure 14.1) is different from that obtained using the original elastic parameters (the intercept is the same but the absolute value of the negative gradient is smaller) but still has the same character. If the (erroneous) V_s in the sand is only 80% of the original value, that is, 1.14 km/s, the respective v becomes 0.305 and the respective AVO curve further deviates from the original. This example shows that v recorded in the well does affect the synthetic-reflection gradient *without changing the response type*. The gradient varies from −0.42 in the first case to −0.32 in the second case and to −0.22 in the third case.

Examine now Figures 14.2 and 14.3, which display laboratory velocity and Poisson's ratio data in over 150 room-dry sand and sandstone samples. Porosity, mineralogy, and velocity in these datasets span wide ranges. Yet, in spite of this tremendous variability, the Poisson's ratio measured on room-dry (air-saturated) sand samples rarely exceeds 0.2. The same is true for dry glass-bead packs in a range of differential pressure between zero and 40 MPa – the measured Poisson's ratio is essentially contained between 0.1 and 0.2 (Figure 14.4).

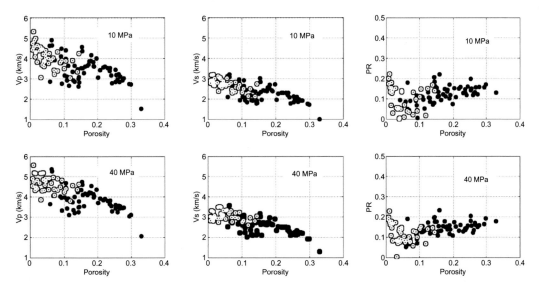

Figure 14.2 Laboratory velocity and Poisson's ratio (PR) versus porosity for sandstone samples at 10 MPa (top) and 40 MPa (bottom) differential pressure. The samples are room-dry. The symbols correspond to the source of the data – filled circles for Han (1986) and circles with black rims for Jizba (1991). The clay content in these samples varies between zero and 50%. The high-porosity low-velocity data point that lies separate from the others is for clean Ottawa sand. After Dvorkin (2008c).

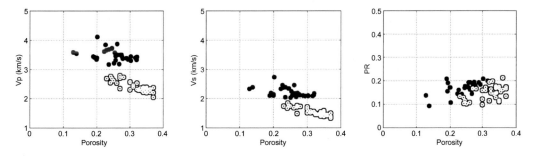

Figure 14.3 Laboratory velocity and Poisson's ratio versus porosity for dry, clean unconsolidated sand samples at 30 MPa differential pressure. The symbols correspond to the source of the data – filled circles for Strandenes (1991) and circles with black rims for Blangy (1992). After Dvorkin (2008c).

These examples confirm an earlier finding by Spencer *et al.* (1994) that the range of v in dry unconsolidated sands is from 0.115 to 0.237 with the mean at 0.187. This result is based on an extensive laboratory dataset that included natural sediment as well as artificial grain packs. The measured v remained small for various grain materials (pure quartz, quartz with clay, corundum, garnet, diamond, calcite, or magnetite).

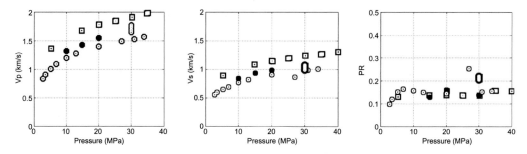

Figure 14.4 Laboratory velocity and Poisson's ratio versus pressure for dry glass-bead packs. The symbols correspond to the source of the data – filled circles for Yin (1992); circles with black rims for Winkler (1979); squares with black rims for Tutuncu (personal communication, 1995); and a group of symbols forming a vertically elongated hollow shape at 30 MPa for Estes (personal communication, 1994). After Dvorkin (2008c).

Effective-medium theories also imply that v in dry porous medium (v_{Dry}) is small. For example, the effective elastic bulk (K_{Dry}) and shear (G_{Dry}) moduli of a random dry pack of identical elastic spheres are (Mavko *et al.*, 2009 and Chapter 2 of this book):

$$K_{Dry} = \frac{n(1-\phi)}{12\pi R} S_N, \quad G_{Dry} = \frac{n(1-\phi)}{20\pi R}\left(S_N + \frac{3}{2}S_T\right), \tag{14.2}$$

where n is the coordination number; ϕ is the porosity of the pack; R is the radius of an individual sphere; and S_N and S_T are the normal and tangential stiffnesses, respectively, for a pair of spheres.

As explained in Chapter 2, these stiffnesses may vary depending on the nature of the grain-to-grain contacts, that is, whether these contacts can only materialize due to externally applied stress or, conversely, by cement that existed at the grain contacts prior to any stress variation. These two extreme situations and their influence on the elastic properties of an aggregate are analyzed in Dvorkin and Nur (1996).

However, no matter what S_N and S_T are, the resulting dry-pack Poisson's ratio derived from Eqs (14.2) is

$$v_{Dry} = \frac{1 - S_T / S_N}{4 + S_T / S_N}, \tag{14.3}$$

which means that v_{Dry} cannot exceed 0.25 (Figure 14.5).

This maximum value of v_{Dry} corresponds to $S_T = 0$, that is, the absence of friction between the particles. At the opposite end of the spectrum is the case of particles without sliding at the contact described by the Hertz–Mindlin model (e.g., Mavko *et al.*, 2009) which provides

$$v_{Dry} = \frac{v_s}{2(5 - 3v_s)}, \tag{14.4}$$

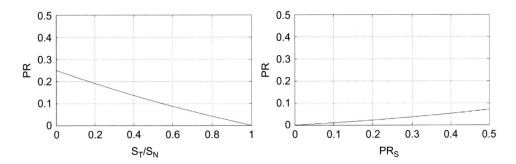

Figure 14.5 Left: The Poisson's ratio of a dry pack of elastic spheres versus the ratio of the contact stiffnesses, according to Eq. (14.3). Right: The Poisson's ratio of a dry pack of elastic spheres with the Hertz–Mindlin contact versus the Poisson's ratio of the grain material, according to Eq. (14.4). After Dvorkin (2008c).

where v_s is the Poisson's ratio of the material of the grains. Figure 14.5 shows that v_{Dry} remains extremely small in the entire range of variation of v_s, a result consistent with the Spencer *et al.* (1994) data.

It is important to remember that the effective-medium equations used here are based on the mean-field approximation that assumes that each and every particle has the same number of contacts and is subject to the same average stress. Sain (2010) shows using numerical particle-scale simulations that this assumption may break, resulting in Poisson's ratio values in dry particle pack larger than those shown in Figure 14.5 (right) but still rarely reaching such extreme values as 0.25 or 0.30. Bachrach and Avseth (2008) attempted to compensate for the apparent mismatch between the mean-field approximation theory and some Poisson's ratio measurements by introducing a correction factor similar to the shear modulus correction factor discussed in Chapter 2.

Another example of using effective-medium theories is provided by the theories of solid with empty (or gas-filled) inclusions. Figure 14.6 shows that, according to the differential effective-medium theory (e.g., Berryman, 1992; Mavko *et al.*, 2009), v_{Dry} in such porous materials is small in a wide porosity range if the inclusions are thin cracks.

Based on these laboratory experiments and theoretical results, one may expect that v measured in the well in gas sand should remain small. The problem as we see it is the apparent inconsistency of v as interpreted from the sonic and dipole velocity data in gas sand intervals: some results indicate that v in such formations can be as small as 0.1 and certainly does not exceed 0.20 (Figure 14.7). Other measurements (Figure 14.8) produce v in gas sand as large as 0.30.

14.2   Physics-based explanations

The question is whether such data should be dismissed as erroneous or taken into account during synthetic seismic generation. In other words, are there situations that

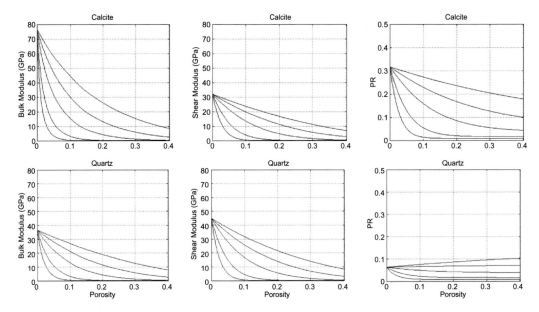

Figure 14.6 From left to right: the bulk modulus, shear modulus, and Poisson's ratio of a solid with empty inclusions. The calculations are according to the differential effective-medium theory. The aspect ratio of the inclusions is 0.20 for the upper curves and 0.01 for the lower curves with values 0.10, 0.05, and 0.02 used in between. The matrix material is pure calcite for the example in the upper row and pure quartz for the example in the lower row. After Dvorkin (2008c).

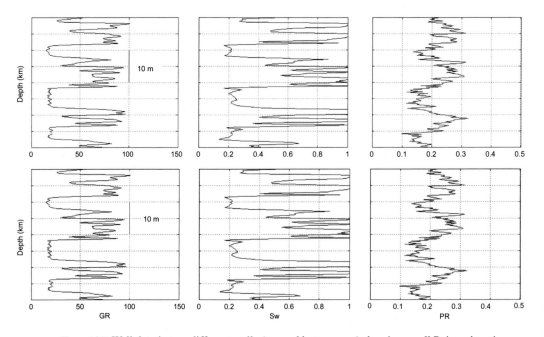

Figure 14.7 Well data in two different wells (top and bottom row) showing small Poisson's ratio (< 0.20) in gas sand (third frame in each row). The gamma-ray (GR) and water saturation are displayed in the first and second frames, respectively. After Dvorkin (2008c).

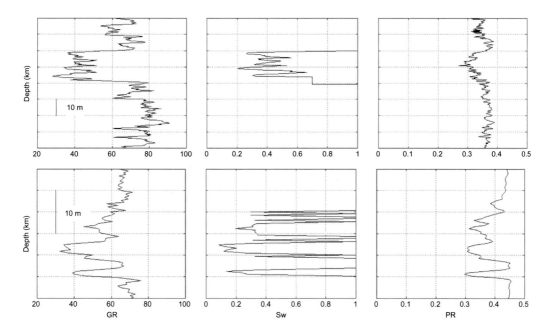

Figure 14.8 Well data in two different wells showing large (> 0.30) Poisson's ratio in gas sand. The display is the same as in Figure 14.7. After Dvorkin (2008c).

may produce relatively high Poisson's ratio in gas sand and still be consistent with the existing experimental evidence? Below we examine three possible explanations for a measured high Poisson's ratio.

Explanation A: Patchy saturation. Consider unconsolidated sand with porosity 0.30 and 5% clay content filled with gas whose bulk modulus is 0.07 GPa and density is 0.21 g/cm^3; and brine whose bulk modulus is 2.75 GPa and density is 1.02 g/cm^3. Let us compute the elastic constants of the dry frame according to the soft-sand model. The elastic constants in the water saturation (S_w) range between zero and 1, and are calculated from those of the dry frame according to Gassmann's (1951) fluid substitution theory (Chapter 2). The effective bulk modulus of the pore fluid (K_f) used in this modeling, a combination of gas and water, is calculated according to the harmonic average

$$K_f = \left(\frac{S_w}{K_w} + \frac{1 - S_w}{K_g} \right)^{-1}, \tag{14.5}$$

where K_w is the bulk modulus of water while K_g is that of gas. The density (ρ_f) is the arithmetic average of those of water (ρ_w) and gas (ρ_g):

$$\rho_f = S_w \rho_w + (1 - S_w)\rho_g. \tag{14.6}$$

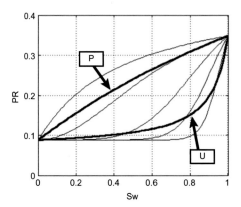

Figure 14.9 Poisson's ratio in sand versus water saturation. The lower bold curve marked "U" is for uniform saturation where the bulk modulus of the water/gas mixture is obtained from Eq. (14.5). The upper bold curve marked "P" is for patchy saturation computed according to Eq. (14.9). The fine curves are according to Eq. (14.8) with parameter e varying from 1 (the upper curve) to 20 (the lower curve) with values 2, 5, and 10 used in between. The larger e the lower the position of the curve due to Eq. (14.8). After Dvorkin (2008c).

The resulting v is plotted versus S_w in Figure 14.9. Its end-member values are about 0.09 at $S_w = 0$ and 0.35 at $S_w = 1$. The transition between these two extremes is rather abrupt: v remains small between zero and 95% water saturation and exceeds 0.25 only in the remaining saturation interval.

Equation (14.5) is appropriate for calculating the effective bulk modulus of a mixture of two fluid phases if they coexist at the pore scale, that is, these phases are *uniformly* distributed in the pore space (see Chapter 2). Domenico (1977) was perhaps the first to observe a departure of laboratory-measured velocity from the velocity predicted using Gassmann's (1951) fluid substitution with the bulk modulus of the pore fluid calculated according to Eq. (14.5). To reconcile his data with the fluid substitution theory, he suggested using an alternative to Eq. (14.5):

$$K_f = S_w K_w + (1 - S_w) K_g. \tag{14.7}$$

Later, Brie *et al.* (1995) noticed a similar effect in well log data and modified Eq. (14.7) as

$$K_f = S_w^e (K_w - K_g) + K_g \tag{14.8}$$

where e is a free parameter. The latter equation becomes Eq. (14.7) for $e = 1$.

Cadoret (1993) discovered that the observed inconsistency between experiment and theory is due to the *patchy* distribution of fluid phases, which means that although S_w may be smaller than 1 in an entire rock sample, the phases are not uniformly distributed in the pore space but rather form fully water-saturated patches that are adjacent to partially saturated regions. A theoretical equation for the extreme case of patchy saturation where the patches are either fully water- or gas-saturated is (Chapter 2)

$$\left(K_P + \frac{4}{3}G\right)^{-1} = S_w \left(K_{S_w=1} + \frac{4}{3}G\right)^{-1} + (1 - S_w)\left(K_{S_w=0} + \frac{4}{3}G\right)^{-1},$$
(14.9)

where K_P is the bulk modulus of porous rock at partial water saturation; G is the rock's shear modulus (independent of S_w); $K_{S_w=1}$ is the bulk modulus of rock at $S_w = 1$; and $K_{S_w=0}$ is the bulk modulus of rock at $S_w = 0$.

The resulting Poisson's ratio calculated as $v = 0.5\left(K_P/G - \frac{2}{3}\right)/\left(K_P/G + \frac{1}{3}\right)$ is plotted versus S_w in Figure 14.9. The dependence of v on S_w is steady in the entire saturation range. Poisson's ratio at patchy saturation exceeds 0.25 for $S_w > 0.5$. Therefore, patchiness in saturation may explain the high v observed in gas wells, especially in the mud filtrate invasion zone. The patchy saturation regime occurs when the wave period is too short to allow pressure equilibration between the two fluids. This results in a higher saturated-rock bulk modulus as well as a higher P-wave velocity at partial saturation. At the same time, it does not affect V_s, thus giving a higher Poisson's ratio.

It still remains necessary to explain whether and why such situations may occur in situ. One such possibility is discussed in Knight *et al.* (1998). It is attributed to a *transient* process of drainage or imbibition where spatial heterogeneity in capillary pressure results in the development of patches at partial saturation. Such a transient process in a well may be due to mud filtrate invasion during drilling. We argue that saturation patchiness perhaps does not exist in virgin formations undisturbed by filtrate invasion. Therefore, a high Poisson's ratio measured at a well due to partial mud filtrate invasion has to be corrected in synthetic seismic modeling used for predicting the seismic response of a virgin formation away from well control.

Of course, considering saturation either homogeneous or patchy is a simplification. Gas and brine are immiscible fluids that form clusters (drops or bubbles). Whether the patchy or homogeneous saturation model applies depends on the characteristic size of these clusters. This size should be compared to the fluid diffusion length, which in turn depends on the frequency of the wave, the permeability of the rock, and the viscosity of the two fluids. Patchy and homogeneous saturation are two end-members which apply in the cases where the size of the cluster is much smaller than the diffusion length and vice versa, respectively. Sonic log and even surface seismic responses may be in either a patchy or homogeneous saturation regime, or between the two end-members, depending on these parameters. Factors that control this behavior are discussed by White (1983), Johnson (2001), and Toms *et al.* (2006) among others. Batzle *et al.* (2006) discuss how patchy limit may even apply at seismic frequencies (for low-permeability rocks such as tight sands).

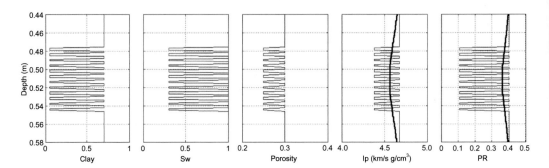

Figure 14.10 A hypothetical interval with 12 thin gas-sand layers. The depth is fictitious. In the impedance and Poisson's ratio frames, the fine curves are for the fine-scale values while bold curves are for the Backus-upscaled results. After Dvorkin (2008c).

Explanation B: Fine layering. Consider now a hypothetical interval containing 12 identical thin gas-sand layers placed in water-saturated shale (Figure 14.10). The clay content in the shale is 70% while in the sand it is 5%. The total porosity in the shale is 0.30 while in the sand it is 0.25. Finally, we assume that the water saturation in the sand is 0.30. The thickness of each gas-sand layer as well as each shale layer in between is 0.3 cm (approximately 0.12 inch).

We use the same water and gas properties as in the previous example and, similarly, calculate the elastic constants in this interval according to the soft-sand model. The P-wave impedance (I_p) in the sand is slightly smaller than in the shale, while v in the sand is significantly smaller than in the shale (Figure 14.10).

These gas-sand layers are thin and below the resolution of conventional sonic and dipole tools. As a result, the elastic constants recorded will be the average of those of sand and shale. To fully ascertain the effect of sub-resolution layering on the reading of the tool, one has to conduct a full-wave-form simulation. A simpler way of quantifying this effect is through the Backus (1962) average (upscaling). The running upscaling window chosen for this example is 10 cm (approximately 4 inches or 0.3 ft).

The resulting Backus-averaged v in the gas-sand sequence may become as large as 0.36, which is significantly different from the local (sub-resolution) v value of about 0.11 in the sand. Although both the impedance and v in the sand sequence remain smaller than in the surrounding shale, their upscaled values change dramatically. This scale-driven change may explain abnormally high v values sometimes observed in gas wells.

Notice that with respect to seismic prospecting this situation is different from that created by patchy saturation. If the layering is sub-resolution at the log scale it will certainly remain sub-resolution at the seismic scale. Therefore, a high Poisson's ratio measured at the well due to thin layering should not be corrected in synthetic seismic modeling.

The physics in this situation is simple: the more shale, the higher Poisson's ratio. If this (layering) mechanism occurs, we should see a reduction in Poisson's ratio with increasing net-to-gross. Arguably, this can be detected from multi-well analysis

Table 14.1 *Data from Yin (1992). Dry Ottawa sand under triaxial loading with constant stresses $P_{xx} = P_{yy} = 0.172$ MPa and varying stress P_{zz}. The measured P-wave velocity in the x, y, and z directions is shown as V_{xx}, V_{yy}, and V_{zz} respectively. The S-wave velocity is listed versus the direction of propagation and polarization as, for example, V_{xy} and V_{xz}, where x is the direction of propagation and y and z are the directions of polarization (see also Figure 14.11). Stress is in MPa and velocity is in km/s.*

P_{zz}	V_{zz}	V_{zx}	V_{zy}	V_{yy}	V_{yx}	V_{yz}	V_{xx}	V_{xy}	V_{xz}
1.03	0.76	0.43	0.42	0.74	0.40	0.43	0.76	0.40	0.42
1.38	0.82	0.44	0.44	0.77	0.42	0.44	0.76	0.43	0.44
1.72	0.87	0.46	0.46	0.79	0.44	0.47	0.79	0.44	0.46
2.07	0.90	0.48	0.48	0.80	0.46	0.47	0.79	0.46	0.49
2.76	0.95	0.52	0.52	0.85	0.48	0.52	0.85	0.48	0.51
3.45	1.01	0.54	0.54	0.87	0.50	0.54	0.87	0.50	0.53

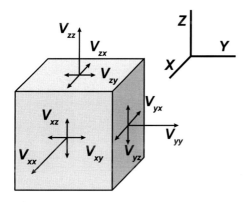

Figure 14.11 Schematic illustration of the compressional- and shear-wave velocity measurements (Yin, 1992) listed in Table 14.1. After Dvorkin (2008c).

(especially if different estimates of net-to-gross, such as gamma-ray, neutron porosity, and NMR, data are available) and thus discriminated from patchy saturation.

Also, in this modeling we assumed that the shale is isotropic. Its possibly strong anisotropy will complicate the situation because the concept of a single Poisson's ratio is not applicable to an anisotropic solid.

Explanation C: Anisotropy. Explanations A and B for the abnormally high Poisson's ratio recorded in gas sand assume isotropy of the sediment. Anisotropy may also contribute to the deviation of the apparent v from expected low values. Consider ultrasonic velocity data obtained in dry Ottawa sand subject to triaxial loading (Yin, 1992, Figure 14.11). The pseudo-shear-wave velocity depends on the shear-wave polarization which, in turn, is a result of stress-induced anisotropy (Table 14.1). In other words, this dataset offers two V_s values for one V_p.

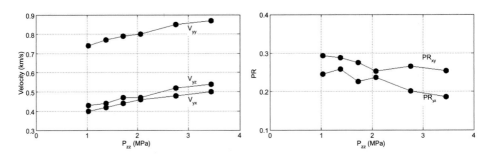

Figure 14.12 Left: Velocity data measured in the y direction under triaxial loading in dry Ottawa sand (Yin, 1992) versus stress in the z direction (Table 14.1). Right: Respective apparent Poisson's ratio. The stress applied in the x and y direction is 0.172 MPa. After Dvorkin (2008c).

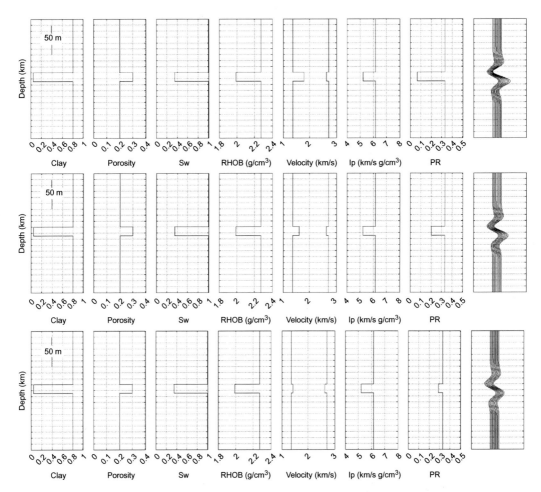

Figure 14.13 Top: The original setup using the soft-sand model. Middle: The S-wave velocity in the sand is reduced by 90%, resulting in Poisson's ratio increase. Bottom: Same but with reducing the S-wave velocity in the sand by 80%. In the synthetic gather trace, the maximum incidence angle is about 45°.

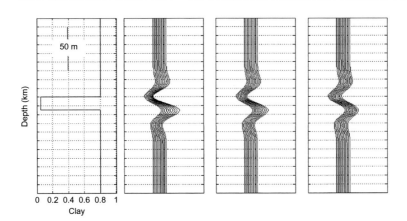

Figure 14.14 Synthetic gathers from Figure 14.13 (top to bottom) place side-by-side (left-to-right).

This situation naturally results in two different apparent values of v as shown in Figure 14.12. While $v_{yz} = 0.5(V_{yy}^2 / V_{yz}^2 - 2) / (V_{yy}^2 / V_{yz}^2 - 1)$ remains fairly small, v calculated using the other S-wave velocity $v_{yx} = 0.5(V_{yy}^2 / V_{yx}^2 - 2) / (V_{yy}^2 / V_{yx}^2 - 1)$ exceeds 0.25 and approaches 0.30. Another example of stress-induced anisotropy is given in Sayers (2002).

This azimuthal anisotropy, whether it is stress-induced or intrinsic, may cause discernible fluctuations in the calculated v in gas-saturated sand. The meaning of this scenario is that the concept of Poisson's ratio should not be used for anisotropic rock. This scenario can be arguably detected by modern logging tools, such as the dipole sonic imager (in cross-dipole regime) or sonic scanner. This effect should not be corrected in synthetic seismic modeling. Rather anisotropy should be incorporated into such modeling.

14.3   But how much does it really matter?

Figure 14.1 indicates that Poisson's ratio may influence the gradient of the P-to-P reflectivity versus angle. Let us explore now how much it will affect the seismic detectability of a gas sand. As before, we will use synthetic seismic modeling of a blocky gas sand with the same fluid parameters as used in Chapter 8. The elastic properties are computed using the soft-sand model (Figure 14.13, top). Then we will gradually reduce the S-wave velocity in the sand by 90 and 80% (Figure 14.13, middle and bottom, respectively).

These synthetic gathers are placed side-by-side in Figure 14.14. We observe that the AVO gradient only slightly changes from the small to intermediate Poisson's ratio cases but its absolute value becomes small as v approaches 0.30 (bottom). The subtle change between the upper and middle cases may be difficult to discriminate in real seismic data. An implication is that we may not have to worry about a precise V_s prediction but rather explore, by means of rock physics modeling and synthetic seismic generation, how much it affects the response and settle for a "good enough" prediction. Of course the definition of "good enough" depends on the application. What is adequate at the exploration stage (e.g., for regional AVO screening) may not be acceptable when conducting seismic impedance inversion for the acoustic and elastic impedances.

15 Seismic wave attenuation

15.1 Background and definitions

Seismic waves attenuate in earth. The number of wavelengths over which the amplitude decreases by a factor of ten is $0.733 \cdot Q$, where Q is the quality factor (see definitions in Mavko *et al.*, 2009 and below). For $V_p = 3$ km/s and frequency $f = 30$ Hz, the wavelength $\lambda = V_p / f = 0.1$ km. Then for $Q = 10$, the amplitude decreases by a factor of 10 as the wave travels $0.733 \cdot Q \cdot \lambda = 0.733$ km.

Such an effect should be accounted for in synthetic seismic modeling at a well or away from well control. This practical method for estimating Q from standard well data is complemented by a more fundamental quest to relate it to rock properties and conditions and their spatial and temporal heterogeneity. The ultimate aspiration is to add attenuation to the arsenal of physics-driven seismic attributes, such as the impedance and V_p / V_s ratio, used for delineating and characterizing reservoirs and monitoring production.

The attenuation coefficient, α, is defined as the exponential decay coefficient of a harmonic wave (Mavko *et al.*, 2009):

$$A(x,t) = A_0 \exp[-\alpha(\omega)x]\exp[i(\omega t - kx)], \tag{15.1}$$

where A is the amplitude of the signal; A_0 is the input-signal amplitude; t is time; x is the spatial coordinate; $\omega = 2\pi f$ is the angular frequency; f is the linear frequency (usually called "frequency"); and k is the wavenumber. The attenuation coefficient is related to the inverse quality factor Q^{-1} as

$$\alpha = Q^{-1}\pi f / V = \pi / (QTV) = \pi / (Q\lambda), \tag{15.2}$$

where V is the phase velocity; T is the period; and λ is the wavelength. By substituting Eq. (15.2) into Eq. (15.1) we obtain

$$\frac{A(x,t)}{A_0} = \exp\left[-\frac{\pi}{Q}\frac{x}{\lambda}\right]\exp[i(\omega t - kx)]. \tag{15.3}$$

To better understand the practical meaning of the quality factor, Q, let us determine over how many wavelengths the amplitude decreases by a factor of 10^n. We find from Eq. (15.3):

$$\exp\left[-\frac{\pi}{Q}\frac{x}{\lambda}\right]=10^{-n} \Rightarrow \frac{x}{\lambda}=n\frac{2.3}{\pi}Q=0.733nQ, \tag{15.4}$$

which means that the required number of wavelengths is $0.733nQ$.

Similarly, the number of wavelengths past which the amplitude decreases by a factor of 2^n is $0.221nQ$. If $Q=5$ ($Q^{-1}=0.2$), the amplitude decreases by a factor of 2 as the wave travels 1.1 wavelengths and by a factor of 10 as the wave travels 3.7 wavelengths. If $Q=10$ ($Q^{-1}=0.1$), the amplitude decreases by a factor of 2 as the wave travels 2.2 wavelengths and by a factor of 10 as it travels 7.3 wavelengths.

Sometimes α is measured in dB per unit length. The conversion coefficient is 8.686:

$$\alpha_{[dB/Length]}=8.686Q^{-1}\pi f / V. \tag{15.5}$$

In an attempt to relate wave attenuation to reservoir properties and conditions, we often have to consider separate attenuation mechanisms described by different mathematical models. For example, elastic waves may attenuate in a dry sandstone frame due to viscoelastic clay present in the frame. If this sandstone is partially saturated, additional attenuation may be due to wave-induced viscous fluid flow. The question is how to add attenuation separately calculated for these two mechanisms to assess the resulting total attenuation.

Let us assume that the first attenuation mechanism acts to reduce the input-signal amplitude by a factor of n, from A_0 to $A_1 = A_0 \exp(-\alpha_1 x)$, over distance x while the second mechanism independently acts to reduce the initial amplitude by a factor of m, from A_0 to $A_2 = \exp(-\alpha_2 x)$, over the same distance. Let us also assume that when acting together, the two mechanisms reduce the initial amplitude over distance x by a factor nm. The resulting amplitude is

$$A_{Sum} = nmA_0 = \frac{A_1}{A_0}\frac{A_2}{A_0}A_0 = A_0 e^{-\alpha_1 x}e^{-\alpha_2 x} = A_0 e^{-(\alpha_1+\alpha_2)x}, \tag{15.6}$$

which means that the attenuation coefficients calculated for separate attenuation mechanisms can be simply added. If we further assume that the phase velocity and dominant frequency are the same for separate mechanisms, we obtain from Eq. (15.2) that the inverse quality factor from separate mechanism adds arithmetically while the quality factor adds harmonically.

What about spatial upscaling? If the amplitude, A_0, of the input signal reduces to $A_1 = A_0 \exp(-\alpha_1 x_1)$ after the wave travels distance x_1 with attenuation coefficient α_1, it further reduces to $A_2 = A_1 \exp(-\alpha_2 x_2)$ after it travels additional distance x_2 with attenuation coefficient α_2. As a result,

$$A_2 = A_0 e^{-(\alpha_1 x_1 + \alpha_2 x_2)} \equiv A_0 e^{-\alpha(x_1 + x_2)}, \tag{15.7}$$

where α is the average (upscaled) attenuation coefficient over distance $x_1 + x_2$. As a result,

$$\alpha = \alpha_1 \frac{x_1}{x_1 + x_2} + \alpha_2 \frac{x_2}{x_1 + x_2}, \tag{15.8}$$

which means that the attenuation coefficient has to be upscaled arithmetically (as $\langle \alpha \rangle$).

Strictly speaking, the inverse quality factor cannot be upscaled arithmetically because $Q^{-1} = \alpha V / \pi f$ and both V and f may change from interval to interval. A correct expression for averaging (upscaling) the inverse quality factor over a long interval is

$$\overline{Q^{-1} \frac{\pi f}{V}} = \left\langle Q^{-1} \frac{\pi f}{V} \right\rangle. \tag{15.9}$$

The average (upscaled) inverse quality factor $\overline{Q^{-1}}$ can be defined through the average velocity $\overline{V}$ and average attenuation coefficient $\overline{\alpha}$ as

$$\overline{Q^{-1}} = \overline{\alpha}\,\overline{V} / \pi f, \tag{15.10}$$

where $\overline{\alpha}$ is the arithmetic average of the attenuation coefficient and $\overline{V}$ is the upscaled velocity, which should be calculated from the Backus (harmonic) average of the elastic modulus $M = \rho V^2$, where ρ is the bulk density:

$$\overline{V} = \sqrt{\overline{M} / \overline{\rho}}, \quad \overline{M} = \langle M^{-1} \rangle^{-1}, \quad \overline{\rho} = \langle \rho \rangle. \tag{15.11}$$

15.2 Attenuation and modulus (velocity) dispersion

If the deformational response of a physical material to a load depends not only on the magnitude of this load but also on the rate of change of the load, the material is called viscoelastic. While in an elastic material the stress tensor, σ_{ij}, is related to the strain tensor, ε_{ij}, by the linear Hooke's law

$$\sigma_{ij} = \lambda \delta_{ij} \varepsilon_{kk} + 2\mu \varepsilon_{ij}, \tag{15.12}$$

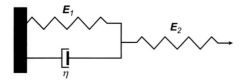

Figure 15.1 A spring/dashpot system with a response described by the SLS constitutive law (Eq. (15.15)).

where λ and μ are Lamé coefficients, such relations in a viscoelastic material are more complicated (Mavko *et al.*, 2009). Some examples of constitutive equations that express these relations are:

$$2\dot{\varepsilon}_{ij} = \dot{\sigma}_{ij} / \mu + \sigma_{ij} / \eta \tag{15.13}$$

for Maxwell's solid;

$$\sigma_{ij} = 2\eta\dot{\varepsilon}_{ij} + 2\mu\varepsilon_{ij} \tag{15.14}$$

for Voigt's solid; and

$$\eta\dot{\sigma}_{ij} + (E_1 + E_2)\sigma_{ij} = E_2(\eta\dot{\varepsilon}_{ij} + E_1\varepsilon_{ij}) \tag{15.15}$$

for the standard linear solid (SLS), where E_1 and E_2 are additional elastic moduli and η is a material constant analogous to viscosity.

If Hooke's law is used to calculate the elastic moduli of a viscoelastic medium, these moduli become complex due to a phase shift between the strain and stress. Of course, the presence of an imaginary part in an expression for these moduli is merely for mathematical convenience.

Physically this simply means that the deformational response of a viscoelastic material to stress is not instantaneous but rather shifted in time. Consider indeed an SLS physical representation by a combination of springs and a dashpot (Figure 15.1). Due to the presence of a viscous dashpot element, the system will react stiffer to fast excitation and softer to slow excitation.

In other words, the apparent effective modulus of the system will be larger for high-frequency excitation than for low-frequency excitation. This effect will translate into the speed of a high-frequency wave being higher than that of a low-frequency wave. It is often called velocity–frequency or modulus–frequency dispersion.

In a viscoelastic medium, the modulus–frequency dispersion and inverse quality factor are linked by the causality Kramers–Kronig relations (Mavko *et al.*, 2009):

$$Q^{-1}(\omega) = \frac{|\omega|}{\pi M_R(\omega)} \int_{-\infty}^{\infty} \frac{M_R(\alpha) - M_R(0)}{\alpha} \frac{d\alpha}{\alpha - \omega},$$

$$M_R(\omega) - M_R(0) = \frac{-\omega}{\pi} \int_{-\infty}^{\infty} \frac{Q(\alpha)M_R(\alpha)}{|\alpha|} \frac{d\alpha}{\alpha - \omega}, \tag{15.16}$$

where ω is the angular frequency and $M_R(\omega)$ is the real part of the complex modulus $M(\omega)$.

Two simple viscoelastic models give examples of linking attenuation to modulus–frequency dispersion (Mavko *et al.*, 2009). According to SLS, the elastic modulus, M, is related to linear frequency, f, as

$$M(f) = \frac{M_0 M_\infty [1 + (f/f_{CR})^2]}{M_\infty + M_0 (f/f_{CR})^2}, \tag{15.17}$$

where M_0 is the low-frequency limit; M_∞ is the high-frequency limit; and f_{CR} is the critical frequency at which the transition occurs from the low-frequency to the high-frequency limit. The corresponding inverse quality factor is

$$Q^{-1}(f) = \frac{(M_\infty - M_0)(f/f_{CR})}{\sqrt{M_0 M_\infty [1 + (f/f_{CR})^2]}}. \tag{15.18}$$

The maximum inverse quality factor is at $f = f_{CR}$:

$$Q_{max}^{-1} = \frac{M_\infty - M_0}{2\sqrt{M_0 M_\infty}}. \tag{15.19}$$

The constant (or nearly constant) Q (CQ) model assumes that the quality factor is constant within a frequency range. Then the maximum inverse quality factor is approximately

$$Q_{max}^{-1} \approx \frac{\pi}{\log(f_1/f_0)} \frac{M_1 - M_0}{2M_0}, \tag{15.20}$$

where M_0 is the modulus at frequency f_0 and M_1 is the modulus at frequency f_1 where both frequency values (f_0 and f_1) are within the constant Q range. It follows from Eq. (15.20) that the modulus changes proportionally to the logarithm of frequency, that is,

$$M = M_0 \left(\frac{2}{\pi Q} \log \frac{f}{f_0} + 1 \right). \tag{15.21}$$

15.3   *Q* data

Consistent and accurate field measurements of Q are rare due to practical difficulties of extracting attenuation from reflection seismic data, cross-well, VSP, and full waveform borehole data.

Reported Q values estimated from seismic events are usually high. Hamilton (1972) reports in situ Q in marine sediments about 30 in wet sand and as high as 100 and even 400 in silt and clay. Kvamme and Havskov (1989) estimate Q about 950 at 10 Hz, while in Lilwall (1988) Q is between 100 and 200 in the upper 3 km of the crust. Leary et al. (1988) use VSP data to find Q exceeding 300 in basement rock at depths below 1.8 km. Quan and Harris (1997) use cross-well tomography to estimate attenuation at the Devine test site in the 200 to 2000 Hz frequency range. Q is between 30 and 50 in soft (V_p between 2.6 and 3.0 km/s) sand/shale sequence and reaches 100 in chalk and limestone. Hackert and Parra (2004) report Q of 33 from a high-resolution 2D seismic data over a Florida carbonate high-porosity aquifer system where V_p is between 2 and 3 km/s and density is about 2 g/cm^3.

A study by Klimentos (1995) is one of the few relevant to hydrocarbon exploration. It reports, based on sonic waveform analysis, that Q falls between 5 and 10 in gas sandstone of about 12% porosity (Q^{-1} between 0.1 and 0.2) while it may easily exceed 100 ($Q^{-1} < 0.01$) in oil- and water-saturated intervals. Attenuation is large in rock with partial gas saturation and small in liquid-fill rock.

The attenuation values obtained in the laboratory are usually larger than those measured in the field. Still, they confirm the fact that large attenuation manifests rock with partial gas saturation. It has been shown that attenuation peaks at small gas saturation (Figure 15.2), an observation potentially useful for monitoring gas depletion in the subsurface. It has been also observed that attenuation is practically non-existent in very dry rock (Figure 15.3).

The extensional-wave data shown in Figures 15.2 and 15.3 come from resonant-bar experiments where there is no stress on the sides of the rock sample. Hence, the wave propagation is governed by the elastic Young's modulus, E, rather than by the compressional modulus, M, so that the speed of the wave with the particle motion along the direction of propagation (there is also a particle motion component normal to this direction) is called the extensional-wave velocity, V_e: $V_e = \sqrt{E / \rho}$. The relation of E to other elastic constants is discussed in Chapter 2 (Eq. (2.2)).

15.4   Modulus dispersion and attenuation at partial saturation

Relaxed and unrelaxed patches. The frequency range of seismic waves used in practical application spans four orders of magnitude, from 10^1 (seismic) to 10^4 (sonic logging) Hz. The pore-scale Biot's and squirt flow attenuation mechanisms (Mavko et al., 2009) are not likely to be engaged at these frequencies. In partially saturated rock, viscoelastic effects and attenuation may arise from the oscillatory liquid cross-flow between fully liquid-saturated patches and the surrounding rock with partial gas saturation. The length scale of these patches can be orders of magnitude larger than the pore scale.

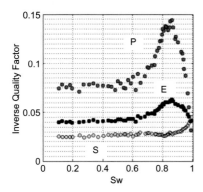

Figure 15.2 The inverse quality factor in Massillon sandstone versus water saturation (after Murphy, 1982). Frequency range is from 300 to 600 Hz. The extensional- and S-wave data (black and light-gray, respectively) are measured while the P-wave inverse quality factor (dark-gray) is calculated from these data (see equation for this calculation in Mavko *et al.*, 2009).

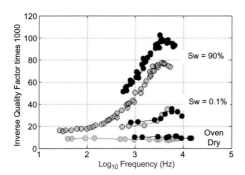

Figure 15.3 The inverse quality factor in Massillon sandstone versus frequency at 90 and 0.1% water saturation as well as in the oven-dry sample (after Murphy, 1982). The extensional-wave data are colored black while the S-wave data are light-gray.

To recognize physical reasons for the existence of patchy saturation, consider a relatively large volume of rock that includes several smaller sand volumes whose clay content and/or grain size vary. Such variations usually dramatically affect permeability and, to some extent, porosity (e.g., Yin, 1992) and, simultaneously, capillary pressure curves and irreducible water saturation.

In a state of capillary equilibrium, capillary pressure is the same for adjacent patches whose irreducible water saturation is different. As a result, at partial saturation, some patches (with large irreducible water saturation) may be fully water-saturated while other patches (with smaller irreducible water saturation) may contain gas (Knight *et al.*, 1998). The whole volume will have patchy liquid distribution.

Visual proof that patches may form in oil/water and air/water systems in the laboratory has been presented by Chatenever and Calhoun (1952) and Cadoret (1993).

Indirect evidence that patches exist in situ has been presented by Brie *et al.* (1995) and Dvorkin *et al.* (1999).

The reaction of rock with patchy saturation to loading due to elastic wave propagation depends on the frequency of the wave. If the frequency is low, that is, the loading is slow, the oscillations of the pore pressure in a fully liquid-saturated patch and partially saturated domains next to it equilibrate. The patch is "relaxed." Conversely, if the frequency is high, that is, the loading of the rock is fast, the resulting oscillatory variations of pore pressure cannot equilibrate between the fully saturated patch and the domain outside. The patch is "unrelaxed." The response of the unrelaxed patch is not influenced by the presence of gas next to it.

The critical size, L, below which the patch is relaxed can be estimated as (see also Chapter 2)

$$L = \sqrt{\frac{1}{f} \frac{k K_w}{\phi \mu}}, \tag{15.22}$$

where k is the permeability; K_w is the bulk modulus of the liquid; ϕ is the porosity; and μ is the dynamic viscosity of the liquid in the patch. Example calculations of the critical size are displayed in Figure 15.4. At 100 Hz and permeability 1 D, this size is about 0.3 m, which means that larger patches will be unrelaxed while smaller patches will be relaxed. For permeability 1 mD, the critical size is about 0.01 m, which means that any patch of a larger size will be unrelaxed at 100 Hz.

Relaxed patches: Low-frequency elastic modulus. If all patches in partially saturated rock are relaxed, which may occur at very low frequency, it is valid to use the concept of the effective pore fluid that is a mixture of liquid and gas throughout the entire volume. The effective bulk modulus of this mixture (K_f) is the harmonic average of the moduli of water (K_w) and gas (K_g):

$$\frac{1}{K_f} = \frac{S_w}{K_w} + \frac{1 - S_w}{K_g}, \tag{15.23}$$

where S_w is water saturation. This type of pore fluid averaging implies "uniform" fluid saturation.

Then the bulk modulus of the partially saturated region, K_{Sat0}, is determined by Gassmann's equation (see also Chapter 2)

$$K_{Sat0} = K_s \frac{\phi K_{Dry} - (1 + \phi) K_f K_{Dry} / K_s + K_f}{(1 - \phi) K_f + \phi K_s - K_f K_{Dry} / K_s}, \tag{15.24}$$

where K_{Dry} is the bulk modulus of the dry frame of the rock; K_s is the bulk modulus of the mineral phase; and ϕ is the total porosity. The shear modulus of the partially

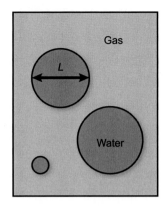 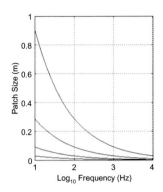

Figure 15.4 Left: Schematic of macroscopic fully saturated patches in a partially saturated reservoir. Right: Critical length versus frequency for a patch saturated with water with the bulk modulus of 2.5 GPa and viscosity 1 cPs, as given by Eq. (15.22). The porosity of the patch is 0.3. The permeability is 1 mD for the lower curve and 1 D for the upper curve with an order of magnitude increment in between.

saturated region, G_{Sat}, is the same as that of the dry-rock frame, G_{Dry}. The compressional modulus at low frequency (M_{Sat0}) is

$$M_{Sat0} = K_{Sat0} + \frac{4}{3} G_{Dry}. \qquad (15.25)$$

For V_p-only fluid substitution (Mavko *et al.*, 1995 and Chapter 2), we can calculate the compressional modulus of saturated rock directly from that of the dry frame (M_{Dry}):

$$M_{Sat0} = M_s \frac{\phi M_{Dry} - (1+\phi) K_f M_{Dry} / M_s + K_f}{(1-\phi) K_f + \phi M_s - K_f M_{Dry} / M_s}, \qquad (15.26)$$

where M_s is the compressional modulus of the mineral phase.

Unrelaxed patches: High-frequency elastic modulus. If the patches in partially saturated rock are unrelaxed, which may occur at high frequency, the concept of the effective pore fluid is not applicable. The bulk modulus of a fully saturated patch (K_P) will be that of the fully liquid-saturated rock:

$$K_p = K_s \frac{\phi K_{Dry} - (1+\phi) K_w K_{Dry} / K_s + K_w}{(1-\phi) K_w + \phi K_s - K_w K_{Dry} / K_s}. \qquad (15.27)$$

If we assume that all liquid in partially saturated rock is concentrated in fully saturated patches and the rest of the rock is filled with gas, the volumetric concentration of the fully saturated patches in the system is S_w. If we assume in addition that the shear modulus is the same for the liquid-saturated and gas-saturated patches, the effective compressional modulus of the partially saturated rock ($M_{Sat\infty}$) is the harmonic average of the compressional moduli of the fully saturated (M_p) and dry ($M_{S_w=0}$) patches (Mavko *et al.*, 2009 and Chapter 2):

$$1 / M_{Sat\infty} = S_w / M_p + (1 - S_w) / M_{S_w=0}, \tag{15.28}$$

or, in terms of the bulk and shear moduli,

$$\frac{1}{K_{Sat\infty} + (4/3)G_{Dry}} = \frac{S_w}{K_p + (4/3)G_{Dry}} + \frac{1 - S_w}{K_{S_w=0} + (4/3)G_{Dry}}, \tag{15.29}$$

where K_p (Eq. (15.27)) and $K_{S_w=0}$ are the bulk moduli of the fully saturated and gas-filled patches, respectively, and

$$K_{S_w=0} = K_s \frac{\phi K_{Dry} - (1 + \phi) K_g K_{Dry} / K_s + K_g}{(1 - \phi) K_g + \phi K_s - K_g K_{Dry} / K_s}. \tag{15.30}$$

Expressions for M_p and $M_{S_w=0}$ using the Mavko *et al.* (1995) V_P-only fluid-substitution equation are

$$M_p = M_s \frac{\phi M_{Dry} - (1 + \phi) K_w M_{Dry} / M_s + K_w}{(1 - \phi) K_w + \phi M_s - K_w M_{Dry} / M_s} \tag{15.31}$$

and

$$M_{Sw=0} = M_s \frac{\phi M_{Dry} - (1 + \phi) K_g M_{Dry} / M_s + K_g}{(1 - \phi) K_g + \phi M_s - K_g M_{Dry} / M_s}, \tag{15.32}$$

respectively.

Attenuation from modulus–frequency dispersion. The difference between the compressional modulus for uniform and patchy fluid saturation is essentially the modulus–frequency dispersion for partially saturated rock. In order to compute the attenuation, we have to assume that the rock is viscoelastic and select a model to describe its behavior. A simple approach (but not necessarily the only possible one) is to use Eq. (15.19) to calculate the maximum inverse quality factor for a given modulus dispersion.

As an example, consider soft sand with porosity 0.30; clay content 5%; and the dry-frame bulk and shear moduli 2.60 and 3.20 GPa, respectively. The bulk moduli of water and gas are 2.64 and 0.04 GPa, respectively. Figure 15.5 displays the low-frequency and high-frequency compressional modulus of this sand versus water saturation as

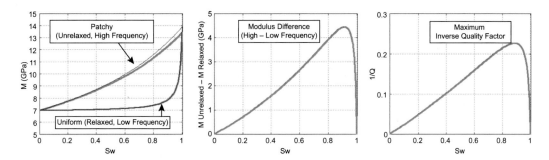

Figure 15.5 Left: Unrelaxed and relaxed compressional modulus versus water saturation for the example in the text. Bold gray curves are for Gassmann's fluid substitution while thin black curves are for the P-wave only fluid substitution. Middle: The difference between the unrelaxed and relaxed compressional modulus (Gassmann's fluid substitution) versus water saturation. Right: The maximum inverse quality factor according to Eq. (15.33).

calculated using these equations. The calculated difference between the low-frequency and high-frequency compressional modulus is zero in dry rock and fully water-saturated rock. The maximum is at about 0.90 water saturation.

Also notice that the difference between the Gassmann fluid substitution results and V_p-only fluid substitution results is very small (Figure 15.5), hence the latter can be used as an accurate approximation.

Finally, we can use the Standard Linear Solid relation between the high- and low-frequency moduli and the maximum inverse quality factor (Eq. (15.19)) to compute the maximum possible $Q_{\max}^{-1}$ as a function of saturation by assuming that the high-frequency compressional modulus is that given by Eq. (15.28) ($M_{Sat\infty}$) while the low-frequency modulus is given by Eq. (15.26) (M_{Sat0}):

$$Q_{P\max}^{-1} = \frac{M_{Sat\infty} - M_{Sat0}}{2\sqrt{M_{Sat\infty} M_{Sat0}}}. \tag{15.33}$$

Because in Eq. (15.33) we use the compressional modulus, the resulting inverse quality factor is related to P-wave propagation. It is plotted versus saturation in Figure 15.5 and reaches maximum at about 90% water saturation, the same as the difference between the unrelaxed and relaxed compressional moduli.

The same approach was used by Cadoret (1993) to estimate attenuation from the velocity measurements performed on a carbonate sample at partial saturation which was uniform in one experiment and patchy in the other. This theoretical estimate was close to the measured attenuation (Figure 15.6).

Let us emphasize that although $Q_{P\max}^{-1}$ varies versus saturation, it is computed as the *maximum inverse quality factor at fixed saturation*: according to Eq. (15.18), it occurs at the critical frequency, f_{CR}. At any other frequency, it will be smaller than given by

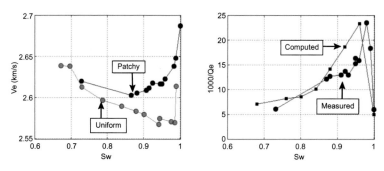

Figure 15.6 Cadoret (1993) data obtained by the resonant-bar technique on a limestone bar. Left: Extensional velocity versus water saturation for patchy (black) and uniform (gray) saturation patterns separately achieved in the same rock sample. Right: The inverse E-wave quality factor measured in the patchy saturation state (black circles) and theoretical estimate of the quality factor based on the difference between the patchy and uniform-state measurement shown on the left (black squares) using the constant Q model (Eq. (15.20)).

Eq. (15.33). Because we generally do not know the critical frequency in natural rock, this $Q_{P\max}^{-1}$ is an estimate rather than a precise value. It provides us with a worst-case scenario for compressional wave amplitude reduction as it travels through rock.

Many authors have attempted to theoretically link attenuation to frequency (see bibliography in Mavko *et al.*, 2009). However, all such theories require inputs (such as the patch size) that simply cannot be obtained in practical situations. This is why we settle here for an upper-bound estimate that can still be useful in evaluating the effects of amplitude reduction on synthetic seismic reflections. Remember also that the energy loss related to wave-induced fluid displacement in rock is just one factor in amplitude reduction. Among others are geometrical spreading and scattering, which have also received extensive treatment in geophysics and material science (see Mavko *et al.*, 2009).

Dvorkin and Uden (2004) theoretically computed attenuation in a gas hydrate well with strong elastic heterogeneity present in the entire interval due to significant stiffening of the rock where hydrate is present. They estimated the scattering attenuation to be about 30% of that due to wave-induced oscillatory cross-flow of the pore fluid (see Chapter 16).

Role of irreducible water saturation. Real gas reservoirs always have irreducible water saturation, S_{wi}. Let us assume that whenever $S_w < S_{wi}$, the pore fluid is distributed within the rock uniformly. This situation was examined in Chapter 2 with the uniform and patchy results for V_p shown in Figure 2.5. The irreducible water saturation strongly depends on the capillary pressure in the rock which, in turn, depends on the pore geometry, porosity, and the surface area of the pore space. The latter two variables affect the permeability. This is why several empirical equations relate permeability, k, to irreducible water saturation and porosity, ϕ, and,

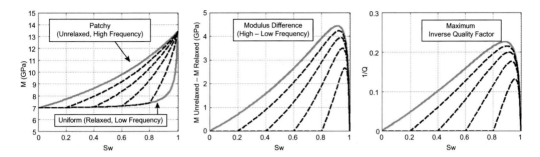

Figure 15.7 Same as Figure 15.5 but with varying irreducible water saturation (dashed curves). The
P-wave only fluid substitution results are not shown.

hence, by using such equations, we can relate S_{wi} to k and ϕ. One such equation is
by Timur (1968):

$$S_{wi} = 11.59(\phi^{1.26} / k^{0.35}) - 0.01, \tag{15.34}$$

where k is in mD and ϕ is expressed as a fraction rather than a percentage.

Let us now use the example shown in Figure 15.5 to explore how the maximum
inverse Q_p varies as a function of S_{wi}. The results shown in Figure 15.7 indicate
that the difference between the patchy and uniform saturation moduli reduces as
S_{wi} increases. So does the maximum inverse quality factor. The implications of this
result for using attenuation to estimate hydrocarbon saturation are important. If the
irreducible water saturation in the reservoir is small and water saturation gradually
increases during production, the curves displayed in Figure 15.7 are relevant and,
hence, one may expect an increase in attenuation as the reservoir is depleted. On the
other hand, if the irreducible water saturation in the reservoir is large and, as a result,
the original reserves are non-commercial (often called the "non-commercial-gas"
situation), the attenuation may not necessarily be an indicator of the *original* water
saturation.

This result underlines the uncertainty of using seismic attenuation as an indicator of hydrocarbon saturation in place. We feel that any seismic attribute, including
attenuation extracted from seismic data, should not be used independently of careful
sedimentological analysis of the subsurface which may provide important evidence
of what reality may stand behind seismic data. Analyses of "non-commercial-gas"
occurrences are few in the literature. One such analysis is given by O'Brien (2004). In
that specific case, the reason for the absence of commercial gas accumulation was the
broken seal above the reservoir (shale cap broken by several permeable faults). This is
a dynamic gas leakage situation as opposed to the static one, due to high irreducible
water saturation. As a result we may expect anomalously large attenuation in the case
analyzed by O'Brien (2004).

15.5   Modulus dispersion and attenuation in wet rock

As proven in the laboratory, the seismic energy is certainly lost in rock with partial water saturation. Still, most of the subsurface is fully wet and seismic waves attenuate as they travel through this rock mass as well. How do we estimate this loss, which is usually smaller than in a reservoir but may have an even larger impact on the seismic amplitude as the travel paths through the wet subsurface are usually much longer than through the reservoirs?

Elastic heterogeneity and macroscopic squirt-flow. Assume that the seismic energy in porous rock with fluid dissipates due to wave-induced oscillatory cross-flow. The viscous-flow friction irreversibly transfers part of the energy into heat. This flow can be especially strong in partially saturated rock where the viscous fluid phase (water) moves in and out of the gas-saturated pore space. Such viscous-friction losses may also occur in wet rock where elastic heterogeneity is present. Deformation due to a stress wave is relatively strong in the softer portion of the rock and weak in the stiffer portion. The spatial heterogeneity in the deformation of the solid frame forces the fluid to flow between the softer and stiffer portions. Such cross-flow may occur at all spatial scales.

Microscopic "squirt-flow" is developed at the sub-millimeter pore scale because a single pore may include compliant crack-like and stiff equi-dimensional parts (Mavko and Jizba, 1991). Macroscopic "squirt-flow" which is more relevant to the seismic prospecting scale may occur due to elastic heterogeneity in the rock frame elastic moduli. This mechanism has recently received a rigorous mathematical treatment by Pride *et al.* (2003) in a "double-porosity" model. However, there is a simpler way of quantifying the effect of macroscopic "squirt-flow" on seismic wave attenuation.

Example. Consider a model rock that is fully water-saturated (wet) and has two parts. One part (80% of the rock volume) is very soft shale with porosity 0.4; clay content 0.8 (the rest is quartz); and the P-wave velocity 1.9 km/s. The other part (the remaining 20%) is clean high-porosity slightly-cemented sand with porosity 0.3 and the P-wave velocity 3.4 km/s. The compressional modulus is 7 GPa in the shale and 25 GPa in the sand. Because of the difference between the compliance of the sand and shale parts, their deformation due to a passing wave is different, which leads to macroscopic "squirt-flow."

At high frequency, there is essentially no cross-flow between sand and shale simply because the flow cannot fully develop during the short cycle of oscillation. The effective elastic modulus of the system is the harmonic (Backus) average (e.g., Mavko *et al.*, 2009) of the moduli of the two parts: $M_\infty = 16$ GPa. At low frequency, the cross-flow can easily develop. In this case, the fluid reacts to the combined deformation of the dry frame of the sand and shale. The dry-frame compressional modulus in the shale is 2 GPa while that in the sand is 20 GPa. The dry-frame modulus of the combined dry frame can perhaps be estimated as the harmonic average of the two: 7

GPa. The arithmetically averaged porosity of the model rock is 0.32. To estimate the effective compressional modulus of the combined dry frame with water we theoretically substitute water into this combined frame. The result is $M_0 = 13$ GPa. Then the maximum P-wave inverse quality factor, Q_{max}^{-1}, according to Eq. (15.19) is large, about 0.1 ($Q = 10$), which translates into a noticeable attenuation coefficient 0.05 dB/m at 50 Hz.

Application to a vertical interval. The above-described averaging technique for attenuation estimate in wet rock can be applied to well log curves by means of a moving-averaging window (Dvorkin and Mavko, 2006). Specifically, in a heterogeneous interval we estimate the average porosity, ϕ_{Eff}, as the arithmetic average of the individual porosities

$$\phi_{Eff} = \langle \phi \rangle; \tag{15.35}$$

and the effective dry-frame compressional modulus, M_{DryEff}, as the Backus average of the individual moduli:

$$M_{DryEff} = \langle M_{Dry}^{-1} \rangle^{-1}. \tag{15.36}$$

The effective saturated-rock compressional modulus at very low frequency can be calculated by applying the V_P-only fluid substitution equation (Mavko *et al.*, 1995) to the domain where the averaging is conducted:

$$M_0 = M_s \frac{\phi_{Eff} M_{DryEff} - (1 + \phi_{Eff}) K_w M_{DryEff} / M_s + K_w}{(1 - \phi_{Eff}) K_w + \phi_{Eff} M_s - K_w M_{DryEff} / M_s}, \tag{15.37}$$

where M_s is the average mineral-phase compressional modulus, assumed the same for all individual parts of the rock. This M_s can be estimated by averaging the mineral-component moduli in the entire interval of the rock by, for example, Hill's (1952) average (Chapter 2).

At high frequency, the individual parts of the interval appear unrelaxed or undrained, meaning that the oscillatory flow simply cannot develop because the period of the oscillation is small and the pore fluid is viscous. In this situation, the saturated-rock compressional moduli of each individual part can be calculated by applying the V_P-only fluid substitution equation individually to each part. The effective saturated-rock compressional modulus of the whole domain is the Backus average of the individual saturated-rock compressional moduli:

$$M_\infty = \left\langle \left(M_s \frac{\phi M_{Dry} - (1 + \phi) K_f M_{Dry} / M_s + K_f}{(1 - \phi) K_f + \phi M_s - K_f M_{Dry} / M_s} \right)^{-1} \right\rangle^{-1}. \tag{15.38}$$

Finally, Q_{max}^{-1} in the interval is calculated in the middle of the moving averaging window from Eq. (15.19) as $0.5(M_\infty - M_0) / \sqrt{M_0 M_\infty}$.

Caveat: Length scale and frequency of oscillatory cross-flow. The proposed theoretical approach to attenuation calculation assumes that wave-induced oscillatory cross-flow can develop between two adjacent elastically different portions of rock within a practical frequency range. The required scale of this elastic heterogeneity is given by Eq. (15.22) and is proportional to the square root of permeability. The permeability in shale is usually very low, of the order of 10^{-2} to 10^{-4} mD or even smaller. As a result, in shale where $k = 10^{-3}$ mD, $\phi = 0.2$, $K_W = 2.7$ GPa, $\mu = 1$ cPs, and for $f = 25$ Hz, we estimate the cross-flow scale, L, as about 1 mm. A measure of elastic heterogeneity at this scale is simply not available from well log data. Therefore, in order for our theory to be valid, we need to assume that heterogeneity discernible at the well log scale reflects much finer heterogeneity (or extends down to a much finer scale).

Equation (15.22) can also be used to estimate the frequency at which oscillatory cross-flow may occur if the smallest scale of elastic heterogeneity is given. If this scale is of the order of 0.5 ft, which is the typical scale of well log measurement, this frequency (f_{CR}) is about 0.001 Hz, which means that if shale is elastically homogeneous within a 0.5 ft span, it will appear "unrelaxed" at any practical frequency.

Therefore, the assumption that the scale of elastic heterogeneity in rock is much smaller than the scale of well measurement but repeats its spatial distribution is required to justify the theory offered here. Still, as will become clear from the following examples, the theory put forward here provides realistic attenuation values (perhaps for a wrong reason?) and, hence, can be used to evaluate its effect on the seismic amplitude.

15.6 Examples

Figure 15.8 shows the inverse quality factor computed using the theory discussed in this chapter for an offshore gas well. The size of the running window was 15 m along the entire depth interval of about 1.5 km. The resulting quality factor in the gas intervals was arithmetically averaged using the same window length. The background attenuation in the wet intervals is small and only peaks where the elastic heterogeneity is discernible (top two peaks in the Q_p^{-1} track). Attenuation is relatively large in the gas intervals where the inverse quality factor approaches 0.10. This appears to be a realistic number, similar to those registered in the laboratory (Figure 15.2).

It is instructive to plot the ratio of the P- to S-wave inverse quality factors (see the S-wave attenuation theory below) versus the V_p / V_s ratio (Figure 15.9). Both attributes point at the presence of gas where both ratios are small. An interesting feature of this plot is that in the gas intervals, the larger the water saturation the larger the Q_p^{-1} / Q_s^{-1}. In light of this result, we need to remind the reader that the attenuation computation in

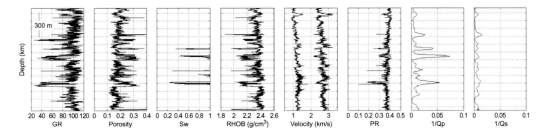

Figure 15.8 An offshore gas well. Depth curve display with the computed *P*- and *S*-wave inverse quality factor shown in the last two tracks. The inverse quality factor curves are smoothed by a running arithmetic average filter.

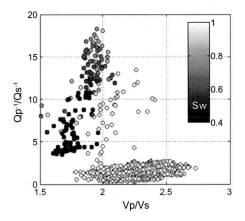

Figure 15.9 The inverse quality factor ratio versus the velocity ratio, color-coded by water saturation. The data come from the example used in Figure 15.8.

this example is model-based, so caution should be exercised in applying the Q_p^{-1} / Q_s^{-1} attribute to quantifying gas saturation in situ.

The next example is for an oil well (Figure 15.10). Because in these wells the hydrocarbon is oil, which is much less compressible than gas (the average value for the bulk modulus of oil is 0.45 GPa compared to 0.13 GPa for gas in the first example), the attenuation rise appears to be driven as much by the elastic heterogeneity as by the presence of hydrocarbons. Also, the Q_p^{-1} value in this case does not exceed 0.02 while in the gas well it was as large as 0.075 in the gas interval.

15.7   Effect of attenuation on seismic traces

To estimate how attenuation affects the seismic amplitude, we will use the same approach as before: modeling a synthetic gather at a well using a ray tracer algorithm where the amplitude is reduced according to Eq. (15.1).

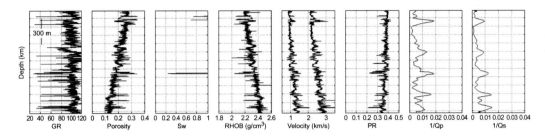

Figure 15.10 Same as Figure 15.8 but for an oil well.

Figure 15.11 shows the resulting seismic gathers computed (a) without taking attenuation into account and (b) using the Q_p^{-1} computed in the well. We also conduct this synthetic modeling for three frequencies of the Ricker wavelet: 30, 45, and 60 Hz. The reflections in the interval become dimmer as the frequency increases. This is because the higher the frequency the shorter the wavelength and, hence, the shorter the distance over which amplitude decrement occurs.

Figure 15.12 shows similar plots for the oil well used in Figure 15.10.

15.8   Approximate theory of S-wave attenuation

Although the S-wave attenuation does not play role in our P-to-P seismic reflection generation, we still describe here a theory for computing the inverse S-wave quality factor (Dvorkin and Mavko, 2006).

Data. Laboratory measurements conducted at ultrasonic frequency on small rock plugs as well as in a lower frequency range using the resonant-bar technique on larger samples indicate that the S-wave inverse quality factor (Q_s^{-1}) is weakly dependent on water saturation and is approximately the same as the inverse P-wave quality factor at full saturation ($Q_s^{-1} \approx Q_p^{-1}$).

Examples include resonant-bar data from Murphy (1982) for Massillon sandstone (Figure 15.2) and ultrasonic data for Vycor glass (Figure 15.13). These Vycor glass data are very close to those presented by Winkler (1979). Prasad (personal communication, 2002) demonstrates the proximity of the P- to S-wave attenuation for an unconsolidated high-porosity sand sample at ultrasonic frequency (Figure 15.13). Lucet (1989) shows that the P-wave attenuation is close to the S-wave attenuation in a limestone sample at ultrasonic frequency. However, in these (Lucet's) experiments, Q_p^{-1} is slightly larger than Q_s^{-1} at low (resonant-bar) frequency.

Reliable field data for Q_p^{-1} and Q_s^{-1} are even more sparse than lab data. Useful results are due to Klimentos (1995), who shows from well log data that the S-wave attenuation is approximately the same as the P-wave attenuation in liquid-saturated sandstone while in gas-saturated intervals the P-wave attenuation is much larger than the S-wave attenuation. Sun *et al.* (2000) compute the P- and S-wave attenuation from

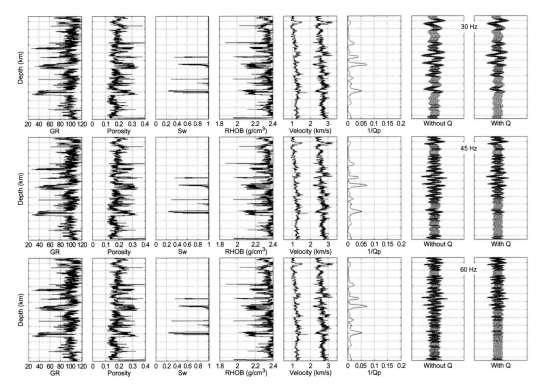

Figure 15.11 Synthetic seismic traces for the well shown in Figure 15.8 computed without accounting for attenuation (7th track) and with attenuation (8th track). From top to bottom: Frequency 30, 45, and 60 Hz.

monopole sonic data. They reported that Q_p^{-1} and Q_s^{-1} are essentially the same in a low-shale-content interval but may be different in the shale.

Attenuation and modulus dispersion. Our first assumption is that the inverse quality factor is related to the modulus–frequency dispersion by a viscoelastic causality relation, such as, for example, for the Standard Linear Solid (Mavko *et al.*, 2009):

$$2Q_p^{-1} = \frac{M_\infty - M_0}{\sqrt{M_0 M_\infty}}, \quad 2Q_s^{-1} = \frac{G_\infty - G_0}{\sqrt{G_0 G_\infty}}, \tag{15.39}$$

where M and G are the compressional and shear moduli, respectively, and the subscripts "∞" and "0" refer, as before, to the high- and low-frequency limits, respectively. We will also assume that the S-wave attenuation is pore-fluid-independent and proceed with our analysis for fully water-saturated porous sediment.

The physical basis for linking the compressional to shear modulus dispersion is the fact that there is a compressional element in shear deformation (pure shear, Figure 15.14). Therefore, if a material includes viscoelastic elements that are responsible for the

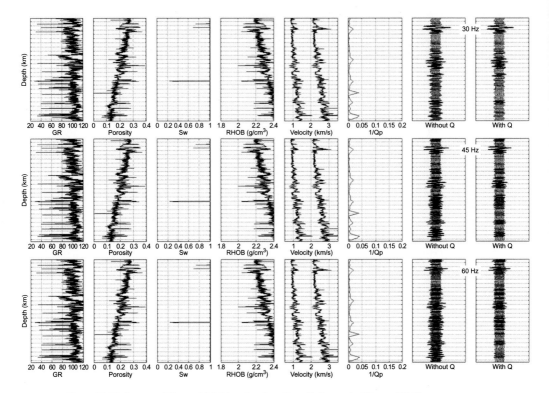

Figure 15.12 Same as Figure 15.11 but for the oil well shown in Figure 15.10.

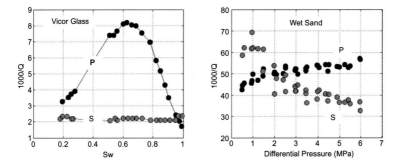

Figure 15.13 Left: Ultrasonic attenuation data in Vycor glass (Murphy, 1982). The inverse quality factor is plotted versus water saturation. Right: Ultrasonic attenuation data in water-saturated unconsolidated sand (Prasad, personal communication, 2002). The inverse quality factor is plotted versus differential pressure.

frequency-stiffening in the compressional-deformation mode, they will contribute to the stiffening in the pure-shear-deformation mode. Mavko and Jizba (1991) use this principle to estimate the contribution of soft crack-like pores containing liquid to the shear-modulus dispersion at ultrasonic frequency at the pore-scale (the microscopic

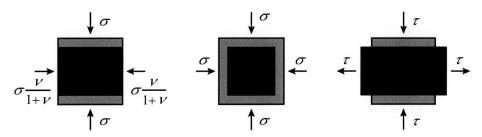

Figure 15.14 From left to right: compressional, bulk, and pure-shear deformation. Gray is the undeformed body while black is the deformed body. The arrows indicate the tractions acting on the body.

squirt-flow). They show that the dispersion of the inverse shear modulus is about 4/15 of that in the inverse bulk modulus.

Here we use the same principle. Specifically, we assume that the reduction in the compressional modulus of *wet* rock between the high-frequency limit and low-frequency limit is due to the *introduction of a hypothetical system of aligned defects (cracks) into the material*. Next, we will adopt Hudsons's theory for cracked media (e.g., Mavko *et al.*, 2009) to quantify these defects. Specifically, the reduction in the compressional modulus in the direction normal to the set of cracks is

$$M_\infty - M_0 = \Delta c_{11}^{\text{Hudson}} \approx \varepsilon \frac{\lambda^2}{\mu} \frac{4(\lambda + 2\mu)}{3(\lambda + \mu)} \equiv \varepsilon \frac{4}{3} \frac{(M - 2G)^2}{G} \frac{M}{M - G}, \tag{15.40}$$

where $\Delta c_{11}^{\text{Hudson}}$ is the change in the anisotropic stiffness component; λ and μ are Lamé constants of the background medium ($\mu \equiv G$); and ε is the crack density – $\varepsilon = 3\phi / (4\pi\alpha)$ – where ϕ is the porosity and α is the aspect ratio. Assuming that $M = \sqrt{M_0 M_\infty}$ we find from Eqs (15.39) and (15.40) that

$$2Q_p^{-1} = \frac{M_\infty - M_0}{\sqrt{M_0 M_\infty}} = \varepsilon \frac{4}{3} \frac{(M - 2G)^2}{G(M - G)} = \varepsilon \frac{4}{3} \frac{(M / G - 2)^2}{(M / G - 1)}. \tag{15.41}$$

The *corresponding change in the shear modulus* for the same set of aligned defects is given by the stiffness component, c_{44}. The change in this component ($\Delta c_{44}^{\text{Hudson}}$) due to the presence of cracks is

$$G_\infty - G_0 = \Delta c_{44}^{\text{Hudson}} \approx \varepsilon \mu \frac{16(\lambda + 2\mu)}{3(3\lambda + 4\mu)} \equiv \varepsilon G \frac{16}{3} \frac{M}{3M - 2G}. \tag{15.42}$$

Assume next that $G = \sqrt{G_0 G_\infty}$. As a result, Eqs (15.39) and (15.42) yield

$$2Q_s^{-1} = \frac{G_\infty - G_0}{\sqrt{G_0 G_\infty}} = \varepsilon \frac{16}{3} \frac{M}{3M - 2G} = \varepsilon \frac{16}{3} \frac{M/G}{3M/G - 2}. \tag{15.43}$$

By combining Eqs (15.41) and (15.43) we find

$$\frac{Q_p^{-1}}{Q_s^{-1}} = \frac{1}{4} \frac{(M/G - 2)^2(3M/G - 2)}{(M/G - 1)(M/G)}, \tag{15.44}$$

where

$$\frac{M}{G} = \frac{2 - 2v}{1 - 2v} = \frac{V_p^2}{V_s^2}, \tag{15.45}$$

and v is, as before, Poisson's ratio.

In another version of the same approach we assume that the same set of defects is now randomly oriented in the material and, hence, does not introduce anisotropy. In this case the reduction in the isotropic shear modulus, $\Delta\mu^{\text{Hudson}}$, is

$$G_\infty - G_0 = \Delta\mu^{\text{Hudson}} \approx \varepsilon \frac{2}{15} \mu \left[\frac{16(\lambda + 2\mu)}{(3\lambda + 4\mu)} + \frac{8(\lambda + 2\mu)}{3(\lambda + \mu)} \right]. \tag{15.46}$$

Hence,

$$2Q_s^{-1} = \frac{G_\infty - G_0}{\sqrt{G_0 G_\infty}} = \varepsilon \frac{16}{15} \left[\frac{2M/G}{(3M/G - 2)} + \frac{M/G}{3(M/G - 1)} \right] \tag{15.47}$$

and, as a result,

$$\frac{Q_p^{-1}}{Q_s^{-1}} = \frac{5}{4} \frac{(M/G - 2)^2}{(M/G - 1)} \left/ \left[\frac{2M/G}{(3M/G - 2)} + \frac{M/G}{3(M/G - 1)} \right] \right. \tag{15.48}$$

Equations (15.44) and (15.48) present two versions for calculating Q_s^{-1} from Q_p^{-1}. It is important to remember that *in these calculations the wet-rock Q_p^{-1} has to be used*, that is, in a hydrocarbon-saturated interval the original fluid has to be substituted for water and Q_p^{-1} calculated afterwards.

The Q_p^{-1} / Q_s^{-1} ratio given by Eqs (15.44) and (15.48) is plotted versus v in Figure 15.15. The two curves differ from each other. However, most importantly, they are close to each other and group around 1 in the Poisson's ratio range between 0.30 and 0.35 which is typical for wet sediment and is consistent with experimental observations.

The Q_s^{-1} computed in an offshore gas well from the wet-rock Q_p^{-1} using Eq. (15.44) is plotted versus depth in Figure 15.8.

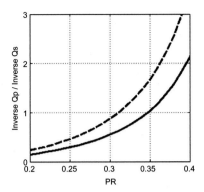

Figure 15.15 *P*- and *S*-wave inverse quality factor ratio versus Poisson's ratio. Solid curve is due to Eq. (15.44) while dashed curve is according to Eq. (15.48).

One of the caveats of this approximate Q_s^{-1} theory (anisotropic case) is that it assumes that the waves propagate normal to the bedding or, more precisely, normal to the hypothetical defects responsible for the modulus dispersion. In more rigorous treatment of the problem, the direction of wave propagation needs to be taken into account or, at least its effects on the errors evaluated.

16 Gas hydrates*

16.1 Background

Gas hydrates are solids where gas molecules are locked inside cage-like structures of hydrogen-bonded water molecules. The physical properties of hydrates are remarkably close to those of pure ice. According to Helgerud (2001), the P- and S-wave velocity in methane hydrate may reach 3.60 and 1.90 km/s, respectively, while its density is 0.910 g/cm³. The corresponding values for ice are 3.89 and 1.97 km/s, and 0.917 g/cm³, respectively. As a result, sediment with hydrate in the pore space, similar to frozen earth, is much more rigid than sediment filled solely by water.

However, unlike ice, methane hydrate can be ignited. A unit volume of hydrate releases about 160 unit volumes of methane (under normal conditions). Also, unlike ice, hydrate can exist at temperatures above 0° C, but not at room conditions – it requires high pore pressure to form and remain stable.

Such stability conditions are abundant on the deep shelf: high pressure is supported by the thick water column, while the temperature remains fairly low (but above 0° C) at depths of several hundred feet below the seafloor because temperature increase with depth starts at a low level, just a few degrees Celsius at the bottom of the ocean. Hydrates also exist onshore below the permafrost which acts to lower temperature at a depth where the hydrostatic pressure is already high. Favorable pressure and temperature are necessary but not sufficient for hydrate generation; its molecular components, water and gas, also have to be available at the same place and time.

Once all these conditions are in place, the elevated rigidity of sediment with hydrate makes it discernible in a seismic reflection volume. Relatively high P-wave impedance of this sediment stands out in the low-impedance background of shallow and unconsolidated deposits. Because the temperature at the seafloor is approximately the same at all depths, at a specific location, hydrates are generated at about the same depth below the seafloor. The top of this stability zone runs parallel to the seafloor, hence its seismic response is a reflector which also runs approximately parallel to the seafloor, the

* Parts of this chapter were modified from work originally published by SEG (Cordon *et al.*, 2006 and Dvorkin and Uden, 2004)

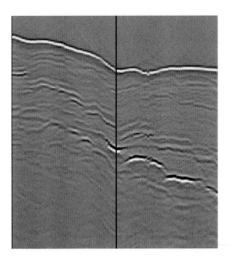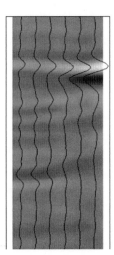

Figure 16.1 Seismic sections at methane hydrate reservoirs. Left: Full-offset stack at the Hydrate Ridge offshore Oregon with the sharp seafloor (top) and hydrate reflections underneath (courtesy Nathan Bangs, UT Austin). Right: A single gather at a hydrate reservoir at another offshore location. Light color indicates positive amplitude (peak) while dark is for negative amplitude (trough).

so-called bottom-simulating reflector (BSR). The free gas, which is sometimes trapped underneath the hydrate-filled host sediment, enhances this reflection, often changes its character, and adds an amplitude-versus-offset (AVO) effect. The presence of gas hydrate (which is a dielectric) in the pore space is also revealed by increased electrical resistivity and, therefore, may be remotely detected by an electromagnetic survey.

BSRs are abundant in the ocean (Figure 16.1). Measurements from dozens of research wells directly confirm that these reflectors are due to methane hydrate. Onshore drilling in Canada, Alaska, and Siberia has also revealed the wide spread of methane hydrate in these Arctic regions. Seasonal hydrate mounds have been visually detected directly on the bottom of the Gulf of Mexico and at other offshore sites. These discoveries indicate that natural hydrates may constitute a gigantic pool of methane: an estimated 7×10^5 Tft^3 of methane is trapped in gas hydrates around the world (Kvenvolden, 1993).

The implications for society are at least threefold: (a) a natural hydrate reservoir can serve as a source of fuel; (b) temporal variations in sea level and earth temperature may act to release methane from destabilized hydrate and vent it into the ocean and atmosphere which, in turn, may affect the atmospheric gas balance; and (c) sediment with hydrate, similar to permafrost, can become a geohazard if disturbed by engineering activity. These factors drive scientific and industrial interest in understanding and quantifying methane hydrate in earth, the latter mostly by means of geophysical remote sensing.

Gas hydrate quantification is, in principle, no different from traditional hydrocarbon reservoir characterization. Similar and well-developed remote sensing techniques can be used, seismic reflection profiling being dominant among them.

Hence, the same question should be addressed: What properties and conditions of a methane hydrate reservoir and surrounding sediment can produce the observed seismic reflections?

16.2 Rock physics models for sediment with gas hydrate

Although the presence of gas hydrate significantly stiffens the rock, the host sediment is usually high-porosity unconsolidated sand located in shallow deposits underneath the seafloor or under permafrost on-shore. Hence, the soft-sand model is likely to relate the elastic properties to the porosity of the host reservoir and gas hydrate saturation of the pore space. Notice that gas hydrates can also exist in mudrock in the deep-water environments where they may be present in cracks and/or form nodules. Here we limit our analysis by examining only the hydrates in sand.

Nevertheless, various modifications of Wyllie's time average equation (Wyllie *et al.*, 1956), which is relevant to fast rock (Mavko *et al.*, 2009), as well as weighted combinations of this model and Wood's (1955) suspension model have found their way into gas hydrate characterization literature (Pearson *et al.*, 1986; Miller *et al.*, 1991; Bangs *et al.*, 1993; Scholl and Hart, 1993; Minshull *et al.*, 1994; Wood *et al.*, 1994; Holbrook *et al.*, 1996; Lee, 2002). Generally, by fine-tuning the input parameters and weights, these equations can be forced to fit a selected dataset. The problem with such fitting is that Wyllie's time average equation is an empirical one and is not based on first physical principles and, hence, should not be used for any kind of sediment. More importantly, combining this model with others is not predictive as it is difficult to establish a rational pattern of adapting free model parameters to site-specific conditions during exploration.

The first breakthrough in the rock physics of gas hydrate was due to Hyndman and Spence (1992). They constructed an empirical relation between porosity and velocity for sediment without gas hydrate and approximated the effect of hydrate presence on sediment velocity by a simple reduction in porosity. By doing so, they effectively assumed that hydrate becomes part of the frame without altering the frame's elastic properties.

Helgerud *et al.* (1999) further developed this idea by using a physics-based effective-medium model to quantify methane hydrate concentration from sonic and checkshot data in a well drilled through a large offshore methane hydrate reservoir at the Outer Blake Ridge in the Atlantic. Sakai (1999) used this model to accurately predict methane hydrate concentration from well log *P*- and *S*-waves as well as VSP data in an onshore gas hydrate well in the Mackenzie Delta in Canada. Ecker *et al.* (2000) used the same model to successfully delineate gas hydrates and map their concentration at the Outer Blake Ridge from seismic interval velocity.

As mentioned earlier, much of the discovered natural methane hydrate is located in clastic and highly unconsolidated reservoirs, either offshore or onshore. We will therefore concentrate on effective-medium models that are relevant to the nature and texture of such sediment.

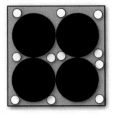

Figure 16.2 Two types of methane hydrate arrangement in the pore space. Left: Hydrate is part of the mineral frame. Right: Hydrate is part of the pore fluid. The mineral grains are black, brine is gray, and hydrate is white.

This model is the soft-sand model where two variants of hydrate distribution are examined: (a) the hydrate mechanically acts as part of the mineral matrix or, conversely, (b) it acts as part of the pore fluid (Figure 16.2)

Assume that the total porosity of sediment is ϕ_t while the hydrate saturation in the pore space is S_h. The remainder of the pore space is filled with water. The porosity available to water is $\phi_t(1-S_h)$. The volume concentration of hydrate, C_h, in a unit volume of rock is $\phi_t S_h$. The total porosity of sediment with hydrate ($\bar{\phi}$) where hydrate is considered part of the mineral phase is

$$\bar{\phi} = \phi_t - C_h = \phi_t(1-S_h) \tag{16.1}$$

and $\bar{\phi}$ becomes ϕ_t for $S_h = 0$ and zero for $S_h = 1$.

The volume fraction of hydrate in the new solid phase that includes the hydrate and the original solid is $C_h / (1-\bar{\phi}) = \phi_t S_h / [1-\phi_t(1-S_h)]$, while the volume fraction of the i-th mineral constituent in this new solid phase is $f_i(1-\phi_t)/(1-\bar{\phi}) = f_i(1-\phi_t)/[1-\phi_t(1-S_h)]$, where f_i is the volume fraction of the i-th mineral in the original mineral frame of the host sediment. The effective elastic moduli and density of the new solid phase material that includes hydrate can be calculated now according to Hill's average (Chapter 2) but using these new volume fractions instead of the original f_i and adding the hydrate. Now the effective elastic properties of the solid frame modified by gas hydrate can be used in the soft-sand model to compute the elastic moduli and velocity in the sediment with gas hydrate.

A different approach to modeling the elastic properties of sediment with hydrate is to assume that the hydrate is suspended in the brine and thus acts to change the bulk modulus of the pore fluid without altering the elastic moduli of the mineral frame. In this case, the total porosity of the mineral frame does not change and remains ϕ_t. The bulk modulus of the pore fluid that is now a mixture of brine and hydrate ($\bar{K}_f$) is the harmonic average of those of hydrate (K_h) and brine (K_f):

$$\bar{K}_f = \left[S_h / K_h + (1-S_h)/K_f \right]^{-1}, \tag{16.2}$$

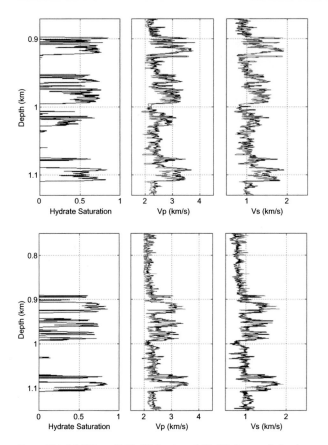

Figure 16.3 Mallik well 2L-38 (top) and 5L-38 (bottom) depth curves. From left to right: hydrate saturation as calculated from resistivity (Cordon *et al.*, 2006); the *P*-wave velocity measured (black) and calculated (gray) using model (a); and the *S*-wave velocity measured (black) and calculated (gray) using model (a).

while its density ($\bar{\rho}_f$) is the arithmetic average of those of hydrate (ρ_h) and brine (ρ_f):

$$\bar{\rho}_f = S_h\rho_h + (1 - S_h)\rho_f. \tag{16.3}$$

In this case, the shear modulus of the sediment with gas hydrate remains unchanged, the same as it was in the wet sediment without hydrate. The bulk modulus is calculated from Gassmann's equation but with $\bar{K}_f$ used instead of K_f. The bulk density, ρ_b, is $\rho_s(1 - \phi_t) + \bar{\rho}_f\phi_t$.

Model (b), although potentially valid, can be rejected in light of existing well data where the *P*- and *S*-wave velocity *both* noticeably increase in the presence of hydrate (if the hydrate is part of the pore fluid, the shear modulus of rock will not be affected by its presence). In contrast, model (a) in which the hydrate is a non-cementing component of the mineral frame matches the data best (Figure 16.3). Essentially, all previous

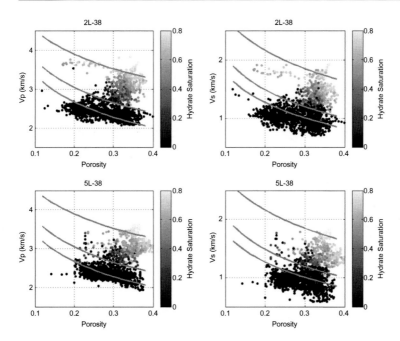

Figure 16.4 Mallik wells 2L-38 (top) and 5L-38 (bottom). *P*- (left) and *S*-wave velocity (right) versus the porosity of the mineral frame (without hydrate) color-coded by hydrate saturation. The model curves are (from top to bottom) for 0.8, 0.4, and zero hydrate saturation in the clean host sand without clay. The data points falling below the zero hydrate saturation curves are from intervals with clay. The well name is listed above each plot.

hydrate-related studies using these effective-medium models to mimic field log and seismic data (Helgerud *et al.*, 1999; Sakai, 1999; Ecker *et al.*, 2000; Dvorkin *et al.* 2003; Dai *et al.*, 2004) arrived at the same conclusion.

Dvorkin and Uden (2004) and Cordon *et al.* (2006) applied model (a) to data from onshore exploratory methane hydrate wells at the Mallik site (Figure 16.3). This model accurately delineates the sands with hydrate from water-saturated sand and shale. Another display of the Mallik modeling results is given in Figure 16.4, where the velocity is plotted versus the porosity of the mineral frame (without methane hydrate) and color-coded by the hydrate saturation of the pore space. The model curves are calculated in a porosity range for wet clean sand without methane hydrate and also for hydrate saturation 0.40 and 0.80. These model curves are consistent with the measured velocity.

The next field example is from the ODP well 995 at the Outer Blake Ridge (Helgerud *et al.*, 1999; Ecker *et al.*, 2000). The sediment at this location is very different from that at the Mallik site: it is predominantly clay with noticeable amounts of calcite and small quantity of quartz. For modeling purposes, Helgerud *et al.* (1999) assumed a uniform mineralogy of 5% quartz, 35% calcite, and 60% clay. The porosity of the mineral frame (without gas hydrate) was that measured on the core material. The hydrate saturation was calculated from resistivity.

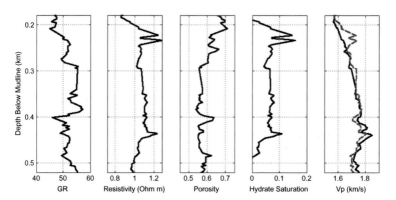

Figure 16.5 Well 995 at the Outer Blake Ridge depth curves. From left to right: GR; resistivity; porosity of the mineral frame without gas hydrate; hydrate saturation; and *P*-wave velocity measured (black) and reproduced by the soft-sand model with hydrate as part of the mineral frame (dashed gray).

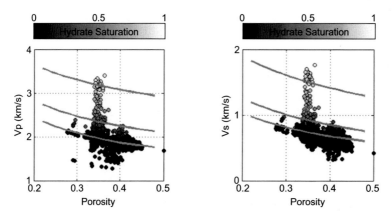

Figure 16.6 Velocity data from two Nankai Trough wells. *P*- (left) and *S*-wave velocity (right) versus porosity of the mineral frame (without hydrate) color-coded by hydrate saturation. The model (a) curves are (from top to bottom) for 0.8, 0.4, and zero hydrate saturation in clean sand with 10% clay. The data points falling below the zero hydrate saturation curves are from intervals with clay.

The sonic velocity is compared to the model (a) predictions in Figure 16.5. The latter reproduces the measurement, except for the upper part where the resistivity indicates the presence of hydrate while the sonic velocity remains low, which points at a possible inconsistency in the well data. This is an example where the model developed for sand actually works in shaly sediment.

The final field example is from yet another depositional environment which is in Nankai Trough offshore Japan where gas hydrate occurs in the sandy parts of the interval (low GR) and is characterized by elevated *P*- and *S*-wave velocity and strong positive reflections. The velocity data are plotted versus the porosity of the mineral frame (without hydrate) and color-coded by hydrate saturation in Figure 16.6. The curves superimposed upon the data are from model (a). In this specific case we assumed that the sand with hydrate contains 10% clay with the rest of the mineral being quartz. Once again, the model curves provide a reasonable match to the data.

As more well data from the methane hydrate environment become available, the rock physics model presented here may have to be fine-tuned, updated, and even changed. This is valid for most of the rock physics "velocity–porosity" models: neither should be taken for granted and applied blindly. The process of data–model feedback may never stop as is typical for other branches of physics.

16.3   Attenuation in sediment with gas hydrate

As discussed in the previous sections of this chapter, elastic-wave data collected in sediments with methane hydrate around the world point to a significant velocity increase due to the presence of the hydrate in the pores. This effect can be easily understood if we recall that gas hydrate is a solid as opposed to brine or gas. By filling the pore space, gas hydrate acts to reduce the porosity available to the pore fluid and thus increase the elastic moduli of the solid frame. *It is difficult to reconcile this effect with observations that the attenuation of elastic waves grows with increasing gas hydrate concentration.*

Indeed, intuitively, one would expect that the stiffer the rock the smaller the relative elastic energy losses per cycle and, therefore, the smaller the attenuation. Measurements in many sediments support this intuition. For example, Klimentos and McCann (1990) show that attenuation increases with increasing porosity and clay content while the velocity behaves in an opposite way. Koesoemadinata and McMechan (2001), who statistically generalized many experimental data, point to the same fact.

Unexpectedly large attenuation in sediments with gas hydrates has recently been observed at different geographical locations, in different depositional environments, and at different frequencies. Guerin *et al.* (1999) presented qualitative evidence of dipole waveform attenuation in the hydrate-bearing sediments in the Outer Blake Ridge. Sakai (1999) noted that the shear-wave VSP signal may be strongly attenuated in a Mallik well within the methane hydrate interval. Wood *et al.* (2000) observed increased attenuation of seismic waves at the same location. Guerin and Goldberg (2002) used monopole and dipole waveforms to quantify compressional- and shear-wave attenuation. They reported a monotonic increase in both with increasing hydrate saturation. Pratt *et al.* (2003) reported an increase in attenuation in the Mallik hydrate reservoir between two methane hydrate wells during cross-hole experiments in the 150 to 500 Hz frequency range.

To explain this effect, Dvorkin and Uden (2004) show that the macroscopic oscillatory flow induced in *wet* rock by a propagating seismic wave may be responsible for the observed strong attenuation in sediment with gas hydrate. This mechanism is discussed in Chapter 15, Section 15.5.

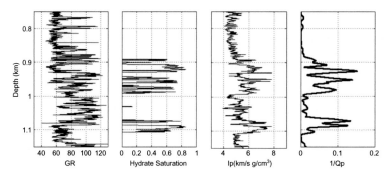

Figure 16.7 Attenuation computed in Mallik well 5L-38 (fourth track).

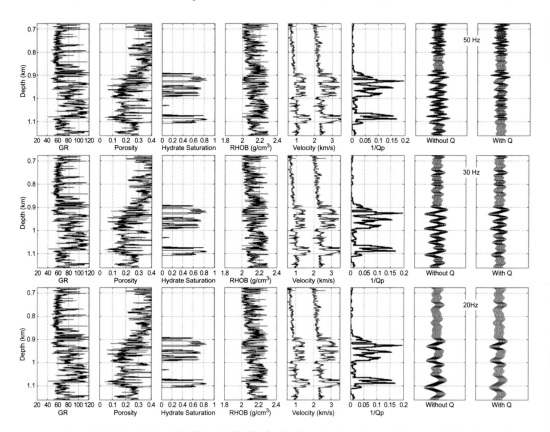

Figure 16.8 Gas hydrate Mallik well 5L-38. Synthetic seismic gathers without and with attenuation (7th and 8th track, respectively) and for frequencies 50, 30, and 20 Hz (top to bottom).

For the purpose of attenuation calculation, the sediment in the interval is considered wet because it does not contain free gas. Then the methane hydrate has to be treated as part of the sediment's frame, according to gas hydrate model (a). Of course, where the hydrate is present, the porosity of this modified frame is smaller than that of the original frame composed of quartz and clay and equals the product of the original

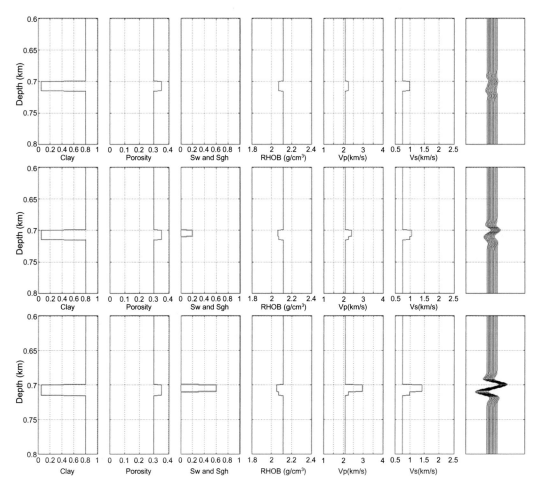

Figure 16.9 Pseudo-well with a wet sand layer whose upper part is occupied by gas hydrate. Gas hydrate saturation is shown in the third track, together with water saturation, as a bold gray curve. Synthetic seismic gather is generated by a 50 Hz Ricker wavelet. The sand contains no free gas.

porosity and one minus methane hydrate saturation (Eq. (16.1)). Also, the effective solid-phase modulus of the modified frame has to include the component due to methane hydrate, as discussed in the previous section. The pore fluid in this modified frame is water.

Although the inverse quality factor, Q_p^{-1}, was small in wet rock in the examples presented in Chapter 15, it may be significantly larger in rock with gas hydrate where the presence of the hydrate acts to increase the stiffness of the rock and thus introduce strong elastic heterogeneity. Dvorkin and Uden (2004) as well as Cordon *et al.* (2006) show that Q_p^{-1} can be as large as 0.10 around intervals with massive gas hydrates (Figure 16.7). This inverse quality factor magnitude is comparable to the empirical equation by Guerin and Goldberg (2002) that links Q_p^{-1} to S_h as

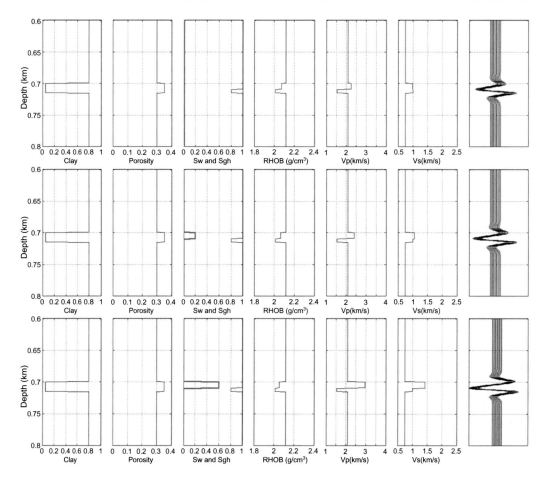

Figure 16.10 Same as Figure 16.8 but with gas sand underneath gas hydrate.

$$Q_p^{-1} = 0.029 + 0.12 S_h. \tag{16.4}$$

Synthetic seismic gathers computed for this well and taking the Q_p^{-1} shown in Figure 16.7 into account are shown in Figure 16.8. The computations are conducted using the Ricker wavelet for three different frequencies: 50, 30, and 20 Hz. As before, the attenuated traces are compared to those computed without attenuation. The amplitude reduction due to attenuation is strong at 50 Hz and becomes hardly detectable at 20 Hz. In geophysics this is called "the low-frequency shadow" (e.g., Castagna *et al.*, 2003, and Ebrom, 2004), meaning that the high-frequency components of the spectrum attenuate at a shorter distance than the low-frequency components and, hence, only low-frequency traces are detectable underneath the high-attenuation interval.

16.4   Pseudo-wells and synthetic seismic in gas hydrates

Figure 16.9 shows synthetic seismic gathers for a well with a blocky wet sand interval with the upper part occupied by methane hydrate as the gas hydrate saturation increases from zero (wet sand) to 0.60.

The sand is quartz and clay. The mineral properties used in this modeling are the same as listed in Table 2.1. The bulk and shear moduli of pure hydrate are 7.7 and 3.2 GPa, respectively, and its density is 0.91 g/cm³. The bulk modulus of water is 2.65 GPa and its density is 1.00 g/cm³. In the following examples we will also fill the bottom part of the sand with gas whose bulk modulus is 0.04 GPa and density is 0.16 g/cm³.

In this example, the small amplitude from the sand without gas hydrate gradually increases as the hydrate saturation increases. The AVO effect in this case is small, as the Poisson's ratio in the shale is about 0.42 decreasing only to 0.35 in the sand with hydrate.

Let us next place small amounts of gas into the sand layer below the gas hydrate. Such an occurrence is common in offshore hydrate deposits, as they are often fed by methane gas seeping from the deeper subsurface. In Figure 16.10 we examine three scenarios with gas saturation 0.20 and gradually increasing gas hydrate saturation in the sand above. In the first scenario, the gas hydrate saturation is zero, so that the gas sand is located beneath the wet sand interval. Although such a situation is unlikely in the setting under examination, we use it here as a hypothetical example.

By introducing gas into the wet sand, we (a) boost the amplitude and (b) increase the AVO effect, as in this case the Poisson's ratio in the gas interval becomes as small as 0.12. The strong amplitude from gas sand without hydrate above it becomes increasingly stronger as the hydrate saturation increases.

Part VI

Rock physics operations directly applied to seismic amplitude and impedance

17 Fluid substitution on seismic amplitude[*]

17.1 Background

All rock physics models require a number of inputs, including porosity and mineralogy. Then there is a host of models to choose from. The procedure of selecting a model based on a training dataset (e.g., well data) has been discussed and utilized in the previous chapters. Once the model is selected, it allows us to explore various "substitution" scenarios, such as lithology, porosity, and reservoir substitution, usually called "what if" perturbations of the original data.

Perhaps the first rock physics substitution equation is that by Gassmann (1951). It is commonly used now to predict the response from a reservoir with a hypothetical "what if" pore fluid from that measured in the well. But can such fluid substitution be conducted directly on the seismic amplitude? Li and Dvorkin (2012) show that it can be done, at least approximately and within a set of assumptions that include establishing a rock physics model relating the elastic properties to porosity and mineralogy. The question posed is whether there are simple recipes that can guide us in predicting a reflection at the shale/gas-sand interface if the reflection at the shale/wet-sand interface is known (and vice versa).

A precursor work addressing essentially the same idea is by Zhou *et al.* (2006). These authors used trends developed from well log data to perform fluid substitution on seismic amplitude and presented a convincing field example to demonstrate the usefulness of the technique. Their assumption was that the reservoirs from down-dip to up-dip are identical except for the pore fluid. Important work by the same group of authors (Ren *et al.*, 2006; Hilterman and Zhou, 2009; Zhou and Hilterman, 2010) addresses the sensitivity of various seismic attributes to water saturation as well as rock type.

The approach used here is rock-physics-driven synthetic seismic forward modeling on a three-layer earth model where a reservoir is sandwiched between two identical shale half-spaces. In the forward-modeling exercise that follows we will cover large

---

[*] This part was modified from work originally published by SEG (Li and Dvorkin, 2012)

ranges of lithology and porosity in both shale and sand, as well as the sand's thickness. By using these synthetic results we directly compare the amplitudes at a reservoir as the fluid varies.

17.2   Primer: model-based reflection between two half-spaces

We start by computing reflections at an interface between two elastic half-spaces where the upper one is occupied by wet shale and the lower is occupied by sand that can be wet or partially gas-saturated. We fix the properties of the water and gas by assuming that their bulk moduli are 2.540 and 0.053 GPa, respectively, and the densities are 0.980 and 0.166 g/cm³, respectively. We also fix water saturation at 40% and compute the bulk modulus of the water/gas mixture as the harmonic average of those of the components while the density of the mixture is the arithmetic average of those of the components. The resulting bulk modulus and density are 0.087 GPa and 0.492 g/cm³, respectively.

Next we randomly select the clay content of the overburden shale between 0.60 and 1.00 and its porosity between 0.10 and 0.30. These parameters in the sand vary between zero and 0.20 and 0.10 and 0.40, respectively. For each pair of the shale and sand properties we compute the P-wave impedance in the sand and shale from either the soft-sand or stiff-sand model (Chapter 2). The normal reflectivity is computed then as the difference between the impedance in the sand minus the impedance in the shale divided by the average of these impedances. This gives us four combinations: soft shale and soft sand; soft shale and stiff sand; stiff shale and soft sand; and stiff shale and stiff sand. Finally, we plot the normal reflectivity from gas sand versus that from wet sand. We repeat these random realizations hundreds of times to fill-in the cross-plots with all possible occurrences.

The normal P-to-P reflection at the shale/gas-sand interface is plotted versus that at the shale/wet-sand interface in Figure 17.1 for the four combinations of the velocity–porosity models. In spite of the large ranges of the rock properties used in this forward modeling, we obtain tight fluid substitution cross-plots.

For example, for the soft-shale/soft-sand combination, if the wet sand is seismically invisible in normal reflections, that is, the intercept is zero, we expect the intercept at the top of the gas sand to vary between about −0.25 and −0.15 (Figure 17.2, left). If the intercept at the wet sand is −0.25, that at the gas sand is expected to vary between −0.45 and −0.35. In reverse, if the intercept at the top of the gas sand is −0.2, that at the top of the wet sand is expected to vary between −0.1 and zero. If the former is −0.4, the latter is about −0.2 (Figure 17.2, right). The same exercise can of course be conducted for a reflection at an angle.

One crucial parameter not examined in this primer is the thickness of the reservoir. The following synthetic seismic modeling will take this effect into account.

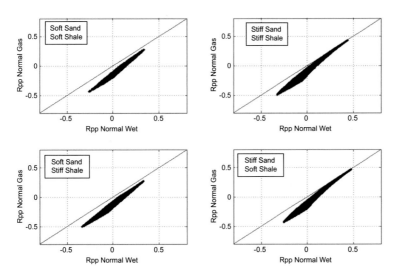

Figure 17.1 Normal reflection at the top of gas sand versus that at the top of wet sand computed as described in the text. The modeling scenarios are listed in the plots. After Li and Dvorkin (2012).

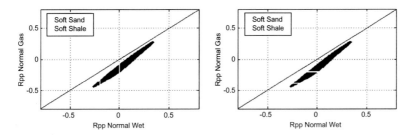

Figure 17.2 Same as the top-left cross-plot in Figure 17.1 but with white bars indicating the ranges of prediction variation as explained in the text. After Li and Dvorkin (2012).

17.3   Model-based effect of thickness

The earth model used here includes a layer of sandstone sandwiched between two infinitely wide and thick identical shale bodies (Figure 17.3). Depending on the depth and the degree of consolidation of rock, at least two rock physics models can be employed to describe how the elastic properties are related to porosity and mineralogy. These are (a) the stiff-sand model representing consolidated rock and (b) the soft-sand model representing unconsolidated and uncemented formations (Chapter 2). We limit our forward modeling by assuming that (a) both shale and sand are consolidated (stiff) and (b) both shale and sand are unconsolidated (soft).

We start with the stiff-sand model and for the purpose of forward modeling fix the properties of the shale surrounding the sand reservoir by assuming that the shale is wet

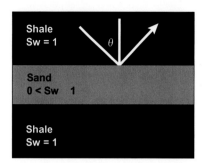

Figure 17.3 Earth model used in synthetic seismic computations.

and its porosity and clay content are 0.15 and 0.70, respectively. The bulk modulus of the formation water is assumed 2.54 GPa while its density is 0.98 g/cm³. The sandstone layer can have variable thickness, porosity, and clay content. The thickness is varied as a fraction of the wavelength, λ, and spans the interval between $\lambda / 2$ and $\lambda / 16$. The porosity of the *stiff sand* varies between 0.10 and 0.30, while its clay content varies between zero and 0.20. This sand can be wet or partially saturated with gas at (assumed) fixed water saturation 40%. The bulk modulus of the gas is assumed 0.053 GPa and its density is 0.166 g/cm³. The effective pore-fluid properties at partial water saturation are computed as the harmonic average for the bulk modulus and the arithmetic average for the density.

In the next set of forward-modeling computations, we use the soft-sand model for both shale and sand. The porosity and mineralogy of the shale remain the same as in the stiff-sand case, however, the porosity of the sand is now varied between 0.15 and 0.35 with its clay content varying in the same interval as used in the stiff-sand case. Also, because unconsolidated rock typically occurs at depths shallower than consolidated rock, we reduce the bulk modulus of the gas to 0.026 GPa and its density to 0.119 g/cm³.

In our forward modeling of the elastic properties of the sand, we randomly and independently vary its porosity and clay content within the above-mentioned ranges. Next, we simulate the synthetic *P*-to-*P* seismic reflection at the top of the reservoir by using the Zoeppritz (1919) equations and convolve the reflectivity series thus produced with the Ricker wavelet of fixed frequency. The angle of incidence in these examples varies from zero to 30°. The AVO attributes, the intercept ($R0$) and gradient (G) are calculated from the synthetic pre-stack seismic data using Shuey's (1985) approximation of the Zoeppritz (1919) equations:

$$R_{pp}(\theta) = R_{pp}(0) + \left[ER_{pp}(0) + \frac{\Delta v}{(1-\overline{v})^2} \right] \sin^2 \theta + \frac{1}{2} \frac{\Delta V_p}{\overline{V}_p} (\tan^2 \theta - \sin^2 \theta), \qquad (17.1)$$

where $R_{pp}(\theta)$ is the P-to-P reflection amplitude at the angle of incidence θ; v is Poisson's ratio; and V_p is the P-wave velocity. Also,

$$E = F - 2(1 - F)\left(\frac{1 - 2\bar{v}}{1 - \bar{v}}\right), \quad F = \frac{\Delta V_p / \bar{V}_p}{\Delta V_p / \bar{V}_p + \Delta \rho / \bar{\rho}}, \tag{17.2}$$

where ρ is the bulk density. In addition,

$$\begin{aligned}
\Delta v &= v_2 - v_1, \quad \bar{v} = (v_2 + v_1)/2; \\
\Delta V_p &= V_{p2} - V_{p1}, \quad \bar{V}_p = (V_{p2} + V_{p1})/2; \\
\Delta \rho &= \rho_2 - \rho_1, \quad \bar{\rho} = (\rho_2 + \rho_1)/2;
\end{aligned} \tag{17.3}$$

where the subscript "1" is for the properties of the upper half-space while "2" is for the lower half-space.

Following Hilterman (1989) we only use the first two terms in Eq. (17.1) as the third term is small at $\theta < 30°$. As a result, the amplitude produced by the Zoeppritz (1919) equations can be fitted by the equation

$$R_{pp}(\theta) = R_{pp}(0) + \left[ER_{pp}(0) + \frac{\Delta v}{(1 - \bar{v})^2} \right] \sin^2 \theta, \tag{17.4}$$

where the first term was used for the intercept, $R0$, while the coefficient in front of $\sin^2 \theta$ in the second term was used for the gradient, G. In most shale-to-sand cases, the gradient is negative as the reflection amplitude decreases with the increasing angle of incidence.

Our objective is to compute these AVO attributes for the two cases: (a) wet reservoir and (b) reservoir with gas. Then we will relate the intercept and gradient at full water saturation to those at partial water saturation and produce best-fit relations between the intercept at the wet reservoir and that at the gas reservoir as well as between the gradient at the wet reservoir and that at the gas reservoir versus varying reservoir thickness, porosity, and the clay content.

Fixed clay content, variable porosity. Figure 17.4 displays the results for the stiff sand and shale for the reservoir's thickness $\lambda/2$, $\lambda/4$, $\lambda/8$, and $\lambda/16$. Figure 17.5 displays the same modeling results but for the intercept. We observe almost linear and fairly tight relations between these attributes computed for wet and gas sand as the sand's porosity varies for each of the selected sand thicknesses.

Moreover, if we superimpose all four graphs from Figure 17.4 on top of each other and do the same with the graphs from Figure 17.5, we still observe that fairly tight linear relations hold for all thicknesses (Figure 17.6).

The same synthetic exercise was conducted for the earth model shown in Figure 17.3, but for the unconsolidated shale and sand whose elastic properties are now related to

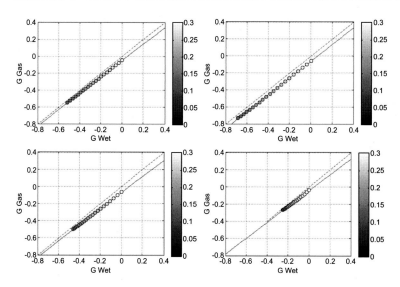

Figure 17.4 The stiff-sand/shale case. The gradient at the top of the gas reservoir versus that at the wet reservoir for a fixed clay content 0.10 and varying porosity (0.10 to 0.30). From left to right and top to bottom: reservoir thickness 1/2, 1/4, 1/8, and 1/16 wavelength. The symbols are color-coded by the porosity of the reservoir. The dashed line is a diagonal. After Li and Dvorkin (2012).

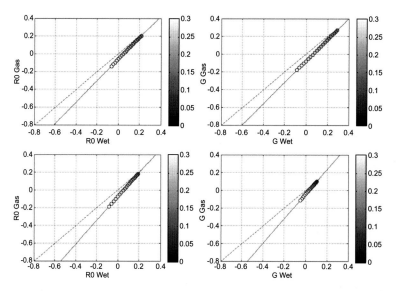

Figure 17.5 The stiff-sand/shale case. Same as Figure 17.4 but for the intercept. After Li and Dvorkin (2012).

porosity and mineralogy by the soft-sand model. The shale's porosity and mineralogy remained the same as in the previous exercise while the porosity of the sand varied between 0.15 and 0.35 with the clay content fixed at 0.10. The summary plots of the gas-sand versus wet-sand gradient and intercept for varying thickness are shown in Figure 17.7.

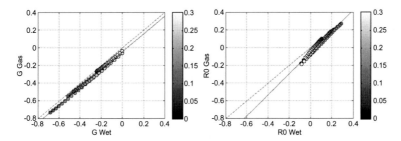

Figure 17.6 The stiff-sand/shale case. Four graphs from Figure 17.4 placed on top of each other (left) and the same for the four graphs from Figure 17.5 (right). The straight lines are the best fit to the modeled points. After Li and Dvorkin (2012).

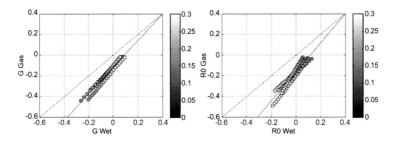

Figure 17.7 Same as Figure 17.6 but for the soft-sand/shale case. The straight lines are the best fit to the modeled points. After Li and Dvorkin (2012).

In this case, the spread of the computed values around the best-fit lines is more prominent than in the stiff-sand case. However, it still may be acceptable bearing in mind the usually sizeable error-bars in real seismic data.

Fixed porosity content, variable clay content. To explore how the clay content in the reservoir affects the transforms between the wet and gas case, we fix the porosity of the stiff-sand reservoir at 0.25 and vary its clay content from zero to 0.20. The cross-plots for varying thickness and the stiff-sand/shale case are shown in Figure 17.8.

The results of exactly the same exercise but for the soft-sand/shale case are shown in Figure 17.9. The linear fit in this case is not as accurate as for the case shown in Figure 17.8.

Variable clay content and variable porosity. Let us now simultaneously vary the porosity and clay content for the stiff-sand/shale case. The properties of the shale remained the same as used in the above examples. The porosity and clay content ranges for the stiff sand also remained the same. We conduct the synthetic seismic modeling for the elastic properties of the sand reservoir corresponding to all combinations of its porosity and clay content and for the wet and gas reservoir cases. We also extend the cases modeled for reservoir thickness $\lambda/2$, $\lambda/4$, $\lambda/8$, and $\lambda/16$.

The results are summarized as the difference between the gradient at the wet and gas reservoir ($\Delta G = G_{Wet} - G_{Gas}$) and the same for the intercept ($\Delta R0 = R0_{Wet} - R0_{Gas}$). Table 17.1 lists the mean, minimum, and maximum ΔG for a fixed reservoir thickness

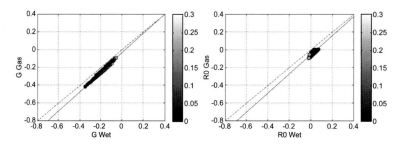

Figure 17.8 Stiff-sand/shale case. Fixed porosity and variable clay content (the color-bar). Display is the same as in Figure 17.6. After Li and Dvorkin (2012).

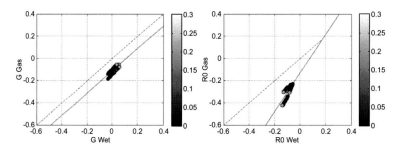

Figure 17.9 Same as Figure 17.8 but for the soft-sand/shale case. After Li and Dvorkin (2012).

and all combinations of its porosity and clay content. The same table lists the mean, minimum, and maximum of $\Delta R0$. The bottom row of this table lists the mean, minimum, and maximum of these differences for *all four thicknesses*.

The same exercise was performed for the soft-sand/shale case with exactly the same parameters as used for the soft-sand/shale case in the previous sections. These results are listed in Table 17.2. These results for both stiff- and soft-sand/shale cases are also plotted in Figure 17.10 as the mean, minimum, and maximum of the gradient and intercept difference versus the inverse thickness which is, as before, expressed as a fraction of the wavelength.

As indicated by Figure 17.10, the error-bar is often unacceptably large. This means that if we do not know the exact values of porosity and clay content in the reservoir, our fluid substitution transform between the reflections at the wet and gas reservoir is not practically acceptable. To *mitigate this situation* let us recall that in many sands the porosity and clay content are related to each other as put forward by the famous Thomas–Stieber model for the porosity in formations with layered shale beds or where structural shale is present (detailed in Mavko *et al.*, 2009). Once such a relation is established (see Chapter 5, Section 5.2), we do not have to independently vary the clay content and porosity: we only have to vary the clay content and assign the porosity values for the pure-sand and pure-shale end-members. The following field case study illustrates this approach.

Table 17.1 *The gradient and intercept differences (minimum, mean, and maximum) between the wet and gas reservoirs, the stiff-sand/shale case, as explained in the text.*

Thickness	Min ΔG	Mean ΔG	Max ΔG	Min $\Delta R0$	Mean $\Delta R0$	Max $\Delta R0$
$\lambda/2$	0.0317	0.0428	0.0498	0.0156	0.0434	0.0853
$\lambda/4$	0.0476	0.0591	0.0669	0.0168	0.0559	0.0928
$\lambda/8$	0.0328	0.0548	0.0682	0.0101	0.0503	0.1221
$\lambda/16$	0.0185	0.0321	0.0416	0.0050	0.0295	0.0794
All	0.0185	0.0472	0.0682	0.0050	0.0448	0.1221

Table 17.2 *Same as Table 17.1 but for the soft-sand/shale case.*

Thickness	Min ΔG	Mean ΔG	Max ΔG	Min $\Delta R0$	Mean $\Delta R0$	Max $\Delta R0$
$\lambda/2$	0.0838	0.1184	0.1516	0.1057	0.1581	0.2076
$\lambda/4$	0.1001	0.1310	0.1919	0.1502	0.1689	0.1821
$\lambda/8$	0.1132	0.1721	0.2238	0.1118	0.2230	0.3093
$\lambda/16$	0.0675	0.1123	0.1405	0.0617	0.1585	0.2750
All	0.0675	0.1334	0.2238	0.0617	0.1771	0.3093

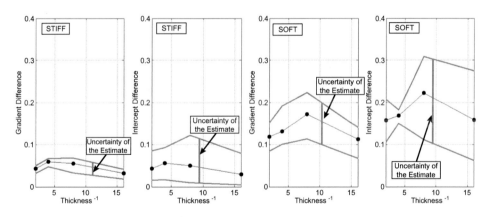

Figure 17.10 The results from Tables 17.1 and 17.2 plotted versus the inverse thickness of the reservoir. The units of the thickness are fractions of the wavelength, meaning that, for example, the value of the inverse thickness 8 corresponds to the thickness 1/8 of the wavelength. The vertical gray bars show the spread around the mean values. After Li and Dvorkin (2012).

17.4 Applying a model-based approach to a case study

The seismic-scale fluid substitution methodology developed on synthetic examples is applied to a full-stack vertical seismic section with a strong negative amplitude at a potential reservoir (Figure 17.11). No pre-stack data or angle stacks are available. A well was drilled on the assumption that this sandstone reservoir contained

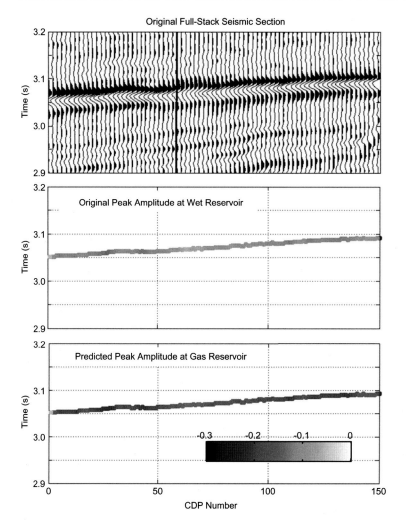

Figure 17.11 Top: Full-stack seismic section under examination (wet reservoir) with the well's position shown by a vertical bar. Middle: The original peak negative amplitude extracted from this section (wet reservoir). Bottom: The peak negative amplitude obtained from that shown in the middle using the third expression in Eq. (17.8). This is the full-stack amplitude expected at a gas reservoir. After Li and Dvorkin (2012).

gas somewhere between 3.70 and 3.80 km. However, in the well, the sand layer appeared to be 100% wet. How will the full-stack seismic response look if gas was present?

The well data in the interval surrounding the potential reservoir are shown in Figure 17.12. The negative amplitude visible in Figure 17.11 (top) at about 3.15 s TWT has apparently been produced by the low-impedance, low-GR sand layer with a higher-impedance layer above, further enhanced by the high-impedance spike between 3.72 and 3.73 km, which could be a carbonate streak or highly-compressed shale.

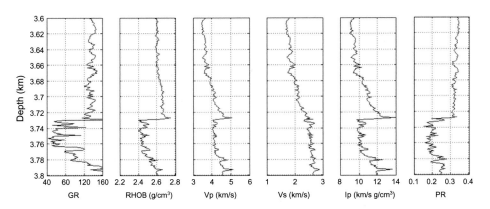

Figure 17.12 Well data for the case study. After Li and Dvorkin (2012).

The total porosity, ϕ, in the well was computed from the bulk density, ρ_b, by assuming that the density of the mineral phase was 2.65 g/cm³ and the density of the fluid was 1.00 g/cm³:

$$\phi = (2.65 - \rho_b) / 1.65. \tag{17.5}$$

The clay content, C, was computed by linearly rescaling the GR curve with the minimum GR value in the section ascribed to pure sand and the maximum GR value ascribed to pure clay.

The velocity–porosity cross-plots for this well are shown in Figure 17.13. Here we observe a clear separation in the velocity–porosity behavior between sand and shale: (a) the sand has a higher porosity, reaching almost 0.15, while the porosity of the shale does not exceed 0.05; and (b) within the overlapping porosity range, the velocity in the sand is higher than in the shale. The latter point becomes even more pronounced when we superimpose model lines on top of the data.

The model we use here is the constant-cement model (the soft-sand model with the coordination number 15); differential pressure 40 MPa; critical porosity 0.40; the bulk modulus of the fluid (water) 2.60 GPa and its density 1.00 g/cm³. The mineralogy is a mixture of quartz and clay with the constants from Table 2.1. The model accurately mimics the data and, therefore, can be used in our fluid substitution workflow.

The third graph in Figure 17.13 is the cross-plot of the total porosity versus clay content. As the clay content increases as the rock transitions from pure sand to pure shale, the porosity decreases. For simplicity, we approximate the observed behavior by a linear trend

$$\phi = (1 - C)\phi_{SS} + C\phi_{SH}; \quad C = (\phi_{SS} - \phi) / (\phi_{SS} - \phi_{SH}), \tag{17.6}$$

where ϕ_{SS} is the porosity of the sand end-member, while ϕ_{SH} is that of the clay (shale) end-member. By selecting $\phi_{SS} = 0.15$ and $\phi_{SH} = 0.03$, we arrive at the clay–porosity relation

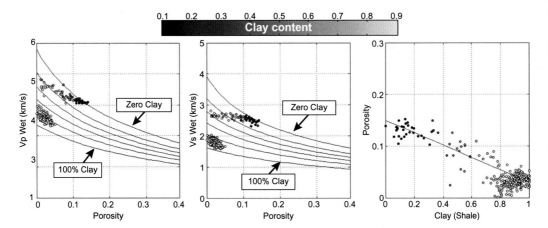

Figure 17.13 First two graphs: Velocity versus porosity from the interval shown in Figure 17.12, color-coded by the clay content. The curves are from the rock physics constant-cement model. The upper curve is for pure-quartz wet rock while the lower curve is for pure-clay wet rock. The curves in between are for gradually increasing clay content (top to bottom) with 20% clay content increment. The third graph is a cross-plot of porosity versus clay content, color-coded by the clay content. The line is an interpolation between the pure-sand and pure-clay points according to Eq. (17.7). After Li and Dvorkin (2012).

$$C = -8.33\phi + 1.25, \quad \phi = 0.12(1.25 - C). \tag{17.7}$$

This equation is plotted as a straight line connecting the pure-sand and pure-shale end-points in Figure 17.13 (right).

Of course, this relation is local and may change in a different depositional setting. Moreover, the C versus ϕ behavior does not have to be linear (see discussion in Mavko *et al.*, 2009).

In order to obtain a seismic-scale wet-to-gas sand transform, we, once again, use the sandwich earth model shown in Figure 17.3. The elastic properties of the wet shale surrounding the sand were computed using the constant-cement model used in Figure 17.13 for porosity 0.04 and clay content 0.80. For the sand body, we varied the porosity between 0.05 and 0.20 and estimated the corresponding clay content from Eq. (17.7). The reflection amplitude was synthetically computed in the same fashion as described earlier in the text for (a) wet sand whose thickness varied between $\lambda / 16$ and $\lambda / 2$; and (b) gas sand with gas saturation 60% and the gas bulk modulus of 0.05 GPa and a density of 0.17 g/cm³.

The resulting cross-plots of the gradient, intercept, and full-stack amplitude (up to 30° angle of incidence) at the top of the reservoir are shown in Figure 17.14 for varying porosity in the sand and its varying thickness. The relations thus produced are close to linear and can be approximated as

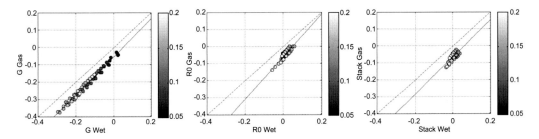

Figure 17.14 Cross-plots of the gradient (top-left), intercept (top-right), and full stack (bottom) at the top of the gas versus wet reservoir produced by synthetic seismic modeling as explained in the text. The symbols are color-coded by the porosity of the reservoir. Different groups of linear trends correspond to different thickness of the reservoir. The straight lines are best-linear-fit approximations for all the points displayed as expressed by Eqs (17.8). Dashed line is a diagonal. After Li and Dvorkin (2012).

$$
\begin{aligned}
G_{Gas} &= 1.0954 G_{Wet} - 0.0559, \quad R^2 = 0.9932; \\
R0_{Gas} &= 1.2447 R0_{Wet} - 0.0617, \quad R^2 = 0.8942; \\
RS_{Gas} &= 1.0945 RS_{Wet} - 0.0674, \quad R^2 = 0.7000;
\end{aligned}
\tag{17.8}
$$

where RS is the full amplitude stack between zero and 30° and all other symbols are the same as used earlier in this chapter.

Now we can apply the wet-to-gas transform expressed by the third expression in Eq. (17.8) to the full-stack field seismic amplitude at the top of the reservoir and, by so doing, translate it into the amplitude at the hypothetical gas reservoir (Figure 17.11, bottom). The ratio of the predicted amplitude to the original amplitude is plotted in Figure 17.15.

17.5   Lessons and conclusions

The question posed in this work was whether it is possible, at least approximately, to transform the seismic amplitude registered at a reservoir that is wet to the amplitude at a hypothetical reservoir that is exactly the same as the wet reservoir but contains gas. In other words, is it possible to conduct fluid substitution directly on seismic data?

We address this question by means of seismic forward modeling conducted on a simple three-layer earth model where a sand is sandwiched between two identical shale bodies. From a number of computational experiments, we find that there are approximately linear relations between the gradient and intercept at a wet reservoir and those at a gas reservoir. These relations appear to be reasonably stable as the porosity and thickness of the reservoir vary. However, if we simultaneously vary both the porosity and clay content in the reservoir, the error-bars of such relations become large. This

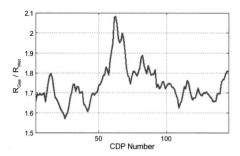

Figure 17.15 The ratio of the full-stack amplitude predicted at the gas reservoir to that at the original wet reservoir. After Li and Dvorkin (2012).

is why we include an additional constraint, that the porosity is inversely related to the clay content. Once such a transform is in place, the uncertainty of the amplitude transform between the wet and gas reservoir is reduced.

This approach is used in our field example where we directly transform the full-stack amplitude at a wet reservoir to that at a gas reservoir under the assumption that the porosity and clay content in the gas reservoir are exactly the same as in the wet reservoir and, moreover, the properties of the shale above and below the reservoir remain the same. This field example does not include explicit validation as we do not have seismic data at a gas reservoir comparable to the wet reservoir under examination. Hence, this case study has to be treated as an example of a change in real seismic data due to replacement of water with gas and using our rock-physics-based technique.

As in any rock-physics-based study, the methodology presented here rests on a large number of assumptions which a diligent student of rock physics has always to explicitly formulate:

(a) the wet and gas reservoirs are exactly the same except for the pore fluid present;
(b) the elastic properties of the shale surrounding the reservoir are known and fixed;
(c) the properties of the water and hydrocarbon (gas) as well as the hydrocarbon saturation are known;
(d) the real earth can be approximated by a simple three-layer model;
(e) the wavelet is known; and
(f) the velocity–porosity and porosity–clay rock physics models are established.

In spite of the limitations forced upon the final result by these assumptions, we feel that they are not much more severe than those used in many traditional seismic modeling studies, meaning that the results presented here are practically usable and useful. However, we encourage the reader to treat this discussion as a method rather than directly taking the equations presented here and applying them to a different field case. Depending on the rock type and real earth geometry, this workflow needs to be implemented on a case-by-case basis with a rigorous rock physics analysis as the basis. Such rock physics models can be established (as shown here) from well data or simply assumed based on the geological circumstances at hand.

Let us emphasize that the direct reflection-based fluid substitution described here also hinges on the fact that we know what fluid is present in the reservoir. In the case under examination, we knew from the well data that the reservoir was wet. What if the well control was absent? In this case one should use a well from an analogue environment to establish the transform and then use a "what if" approach to generate a gas-reservoir amplitude by assuming that the original reservoir is wet, and vice versa. Then, the resulting amplitudes should be compared to those from real seismic data to ascertain the risk of drilling and completing the well.

17.6   Practical application

How to use this method where well data are not available? To construct an appropriate earth model, we will need to make assumptions about the effects of compaction on the properties of the reservoir and non-reservoir rock, as discussed in Chapters 5 and 11. We also need to make assumptions about the mineralogy, based, for example, on the assumed sedimentology of the subsurface feature under examination. Then, based on the plausible ranges of porosity and clay content, as well as the rock physics model selected, we can create site-specific transforms, similar to those shown in Figure 17.14 and expressed by Eqs (17.8). Finally, these transforms will be applied to the seismic attributes registered at the seismic event of interest to answer such questions as what these attributes will be at a gas or oil reservoir if we assume that the reservoir is wet and vice versa. Such "what if" modeling can help in better identifying and risking a potential target. Also remember that an important step in practical applications is amplitude calibration (see Hilterman, 2001).

18 Rock physics and seismically derived impedance*

Seismic impedance inversion is a commonly used procedure in practical geophysics. Many commercial packages are aimed at translating seismic traces that react to the contrast of the elastic properties into absolute impedance volumes. The goal of this operation is obvious: because the absolute rather than relative values of the elastic properties can be related to porosity, mineralogy, texture, and fluid via rock physics transforms, such transforms can, in principle, be applied to the seismically derived impedance volume to ascertain the rock properties in the subsurface. For a detailed review of methods and practices, we refer the reader to an encompassing treatise on impedance inversion by Latimer (2011) where a multitude of authors and their publications are listed.

Here we present a case study by Dvorkin and Alkhater (2004) based on simple rock-physics-based logic to delineate porosity and fluid in an impedance section. The reservoir under examination consists of relatively soft sands. As a result, the acoustic impedance of the gas-saturated sand is much smaller than that of the oil- and water-saturated sand. This large impedance difference allows us to identify the pore fluid only from the P-wave data, without using offset information. As a result, we map both pore fluid and porosity by using impedance inversion applied to stacked seismic data.

The vertical section of the P-wave impedance obtained by inversion at a North Sea location is shown in Figure 18.1 (top). The gas/oil contact is located at the flat impedance feature at about 1700 ms TWT. The relief of the reservoir abruptly changes in the middle of the section, which is typical for the step-like offshore sediment geometry in the field under examination.

The two wells used in this impedance inversion are located at CDP gathers 3 and 190, respectively. Both wells that bound this seismic line are producers and penetrate an extensive gas cap and the underlying oil and water intervals (see well log curves in Figure 18.2). The pore-fluid system in the reservoir includes gas, oil, and water. We assume that there is only gas and water above the gas/oil contact and only oil and water between the gas/oil and oil/water contacts. In other words, gas and oil do not coexist in the reservoir. Water saturation is small in the thin oil leg located above wet rock. In spite of GR variations present in the intervals, the rock is essentially clean sand with a small shale content.

---

* This part was modified from work originally published by EAEG (Dvorkin and Alkhater, 2004)

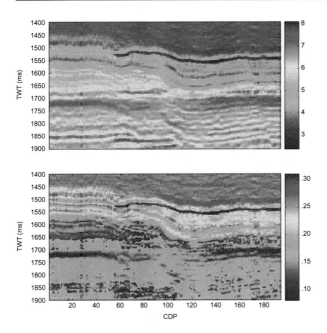

Figure 18.1 Top: Seismically derived impedance section. Bottom: Porosity section obtained by applying Eqs (18.1) and (18.2) to the impedance section. After Dvorkin and Alkhater (2004). The units of the impedance are km/s g/cm³. For color version, see plates section.

The seismically derived impedance shown in Figure 18.1 (top) closely matches the vertical impedance variation visible in both wells: it drastically increases across the flat spot at about 1700 ms TWT which manifests the gas/oil contact (GOC). Compared to this contrast, further impedance variations below GOC are minor.

Figure 18.3 shows the impedance–porosity cross-plots based on the well data. Instead of conducting thorough rock physics diagnostics, we use the "good enough" principle which allows us to derive an approximate simple rule: the impedance in the gas cap predominantly falls below 6 km/s g/cm³ while the impedance of the oil- and water-saturated intervals falls above the 6 km/s g/cm³ mark. This threshold can serve now as the fluid denominator in the impedance section shown in Figure 18.1.

We can now separately concentrate on the data from the gas and oil legs in both wells. An approximate polynomial fit to these trends provides us with the following equations that relate porosity to impedance:

$$\phi \approx 0.775 - 0.151I_p + 0.009I_p^2 \tag{18.1}$$

for $I_p < 6$ km/s g/cm³ (in the gas leg) and

$$\phi \approx 1.100 - 0.185I_p + 0.009I_p^2 \tag{18.2}$$

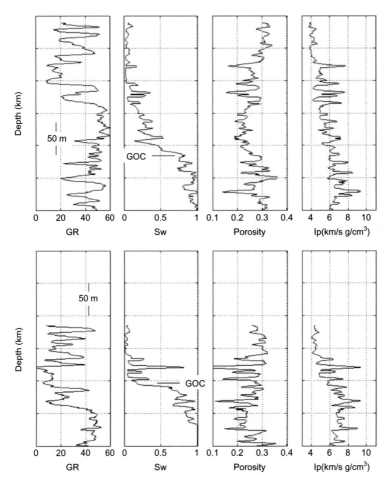

Figure 18.2 Smoothed depth curves in Well 1 (top) and Well 2 (bottom). The horizontal bar marked GOC is for the gas/oil contact.

for $I_p > 6$ km/s g/cm³ (in the liquid leg). The respective curves are superimposed upon the well data in Figure 18.3. The same plots but for the well data upscaled using a 10-point running averaging window are shown in Figure 18.4.

Finally, by using Eq. (18.1) above the flat spot in the impedance section and Eq. (18.2) below it, we can map seismically derived porosity (Figure 18.1, bottom). In essence, by looking at well data and understanding their meaning, we have simultaneously resolved the seismically derived impedance section for porosity and fluid.

Notice that the porosity section is qualitatively different from the impedance section: it does not reflect the sharp horizontal impedance contrast at the flat gas/oil contact reflector. This is because before using the impedance–porosity transforms, we identified the pore fluid and then applied the transforms separately to the gas and liquid legs.

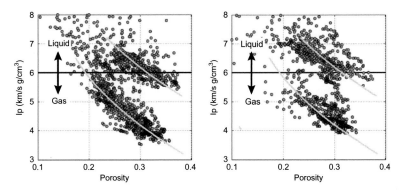

Figure 18.3 Impedance versus porosity in Well 1 (left) and Well 2 (right). Gray curves are according to Eqs (18.1) and (18.2).

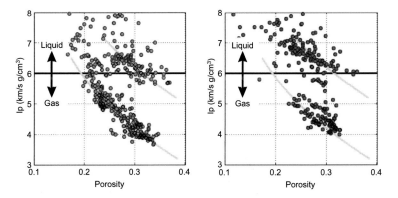

Figure 18.4 Same as Figure 18.3 but for upscaled well data.

This example of applying rock physics to seismic data does not undermine the need for detailed rock physics diagnostics. Only by understanding the rock/fluid system can we start cutting corners to arrive at a simple approximate solution. Moreover, the example discussed in this chapter is site-specific and Eqs (18.1) and (18.2) should not be used in any other field and basin.

Another way of translating seismic data into rock properties and conditions is inverse rock physics modeling which combines rock physics and inverse theory (Grana and Della Rossa, 2010). This work uses a probabilistic approach as required by the general non-uniqueness of translating the amplitude into rock properties (Chapter 1).

There is a host of issues related to applying rock physics to seismic data; one of them is the scale of measurements. One should not necessarily expect the trends derived from the well data that samples the earth at approximately 1 ft scale to be applicable at the seismic scale that may be as large as 100 ft (see Dvorkin and Cooper, 2005; Dvorkin and Uden, 2006; Dvorkin, 2008d).

The resolution of reservoir properties produced by integrating rock physics and seismic data still does not exceed seismic resolution. As a result, one should not expect that, without additional stringent assumptions, reservoir properties can be derived at sub-seismic resolution. By using the deterministic approach described here, we can only determine average properties, such as average porosity.

Part VII

Evolving methods

19 Computational rock physics*

19.1 Third source of controlled experimental data

Laboratory experiments and well data serve as main sources of controlled experimental data where a number of physical properties are measured on the same samples at varying conditions, such as saturation and pressure. Chapter 2 discusses how these data are used to derive theoretical models as well as establish the relevance of these models to rock types.

Computational rock physics, also called digital rock physics or DRP, is the third such source. The principle of this technique is "image and compute": image the pore structure of rock and computationally simulate various physical processes in this space, including single-phase viscous fluid flow for absolute permeability; multiphase flow for relative permeability; electrical flow for resistivity; and loading and stress computation for the elastic properties.

The principle of DRP is simple but its implementation is not. It requires at least three main steps: imaging; image processing and segmentation; and physical property simulation.

Three-dimensional *imaging* of a rock sample is usually performed in a CT scanning machine by rotating the sample relative to an X-ray source. The actual 3D geometry is reconstructed tomographically from these raw data and the image appears in shades of gray. The brightness of a voxel in such a 3D image is directly affected by the effective atomic number of the material and is approximately proportional to its density. For example, dense pyrite will appear bright while less dense quartz will appear light gray. The empty pore space will be black and parts of it filled with, for example, water or bitumen will be dark gray. To image very small features present in shale or micrite in carbonates, even the sharpest CT resolution may not be enough. A different technique, the so-called FIB-SEM, is used where the focused ion beam gradually shaves off thin slices of the sample and the exposed 2D surface is imaged (photographed) by the scanning electron microscope to produce a stack of closely spaced 2D images. This technique allows for resolution as fine as 5 to 10 nm. In imaging, there is always a conflict

* This part was modified from work originally published by SEG (Dvorkin *et al.*, 2011)

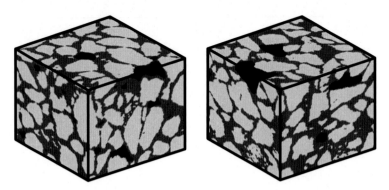

Figure 19.1 Segmented images of different parts of the same bitumen sand sample. The grains are white, bitumen is gray, and air pockets are black. Courtesy Cenovus and Ingrain, Inc.

between the resolution and field of view: the finer the former the smaller the latter. In fact, the size of a typical sandstone or carbonate sample imaged is of the order of millimeters. It is even smaller for shale. Can we learn anything at all about the reservoir-scale behavior of rock from samples this small? We will address this question in one of the following sections.

An image thus acquired has to be *segmented* to separate the pores from the grains and, within the grains, identify the mineral composition. The latter is important in stress-loading computational experiments, as different minerals have different elastic properties and, hence, deform differently from each other. The segmentation process can be as complex as the image processing field itself. It provides critical input to the computational simulation of the physical processes in the pore space, as its connectivity and other geometrical properties depend on how the pores are separated from the grains. Yet, in spite of these difficulties, segmentation has been successfully implemented as indicated by the results of computational experiments that provide verifiable physical properties of rock (see examples later in this chapter). An example of a segmented image of bitumen sand is shown in Figure 19.1 with the grains, bitumen, and air pockets separated from each other.

There are many *computational engines* for process simulation. Absolute and relative permeability may come from simulating single-phase and multiphase flow using the lattice–Boltzmann method (LBM) (e.g., Tolke *et al.*, 2010). For the electrical and elastic simulations, the finite element method (FEM) (e.g., Garboczi and Day, 1995) can be used. In all simulations, appropriate local physical constants are assigned to each voxel (the mineral bulk and shear moduli, conductivity, viscosity, wettability angles, and interfacial tension). In flow simulations, these parameters must come from laboratory tests or be computed for relevant reservoir conditions from tables and equations available. On the one hand, this requirement appears restrictive. On the other hand, this flexibility can be considered advantageous since we can rapidly compute and address various scenarios relevant to the life span of a reservoir, which is extremely difficult if not impossible in the physical laboratory. In electrical flow simulations, assigning local conductivities is fairly straightforward for rock filled with conductive brine and made of virtually nonconductive minerals, such as quartz or calcite. It becomes challenging but manageable where conductive clay, pyrite, or a conductive microporous element (such as micrite) is present. A major advantage of computational rock physics is that all experiments are conducted on the same digital object which can be stored for as long as needed and revisited as new questions and demands arise.

A significant number of publications have been dedicated to DRP. We refer the reader to some of them: Bosl *et al.* (1998), Keehm *et al.* (2001), Arns *et al.* (2002), Øren and Bakke (2003), Knackstedt *et al.* (2003), Dvorkin *et al.* (2008), Dvorkin (2009), Sharp *et al.* (2009), Tolke *et al.* (2010), and Dvorkin *et al.* (2011).

19.2   Scale of experiment and trends

The scale of controlled rock physics experiments ranges from sub-millimeter in DPR to a few centimeters in the laboratory and tens of centimeters in the well. The scale of seismic investigation is tens or even hundreds of feet. How relevant are these small-scale experiments to quantifying the properties of large objects in the subsurface? Specifically, how accurate is DRP and how can its results be validated?

An intuitive answer is in matching the DRP-acquired property of a microscopic test to that measured in the laboratory or well on a larger host sample. The properties often do not match (Dvorkin and Nur, 2009). They do not have to match: because natural rock is heterogeneous at all scales, the properties of a small subsample used in imaging and computing do not have to match the properties of its centimeter-sized host, no more so than the properties of a laboratory sample have to match those inferred for a 30.5-m (100-ft)-sized subsurface object from remote sensing.

To assess the validity and, eventually, apply DRP results to field-scale problems, a different approach can be used: instead of comparing a property obtained at different scales, we should compare *trends* formed by pairs of data points, such as between porosity and permeability, electrical formation factor and porosity, and porosity and the elastic moduli. Kameda and Dvorkin (2004) show that a realistic permeability–porosity trend can be obtained from just a single digital sample by subdividing it into a number of smaller subsamples and computing the aforementioned properties on each of the subsamples. By such subsampling, we can widen the porosity range, as porosity is heterogeneous even within a millimeter-sized rock test. It is remarkable that often the permeability computed on the subsamples, if plotted versus porosity, forms a trend and the properties computed on the whole sample fall on this trend.

The schematics of subsampling are shown in Figure 19.2. Actual subsampling applied to millimeter-sized samples of Berea sandstone and heavy oil sand is shown in Figure 19.3 (Dvorkin *et al.*, 2011).

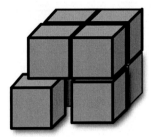

Figure 19.2 Subsampling of a sample into eight even-sized fragments.

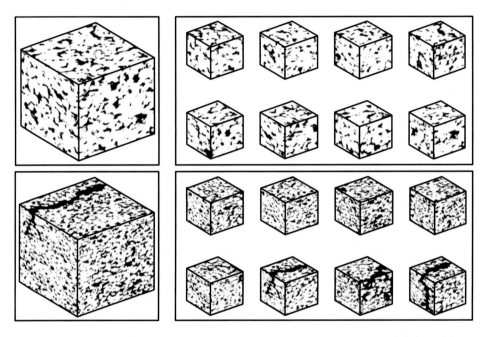

Figure 19.3 Subsampling of Berea sandstone (top left) and heavy oil sand (bottom left) into eight subsamples. After Dvorkin *et al.* (2011).

Both porosity and permeability computed on the eight subsamples cover relatively large ranges. In the whole Berea sample the porosity is 0.15, while the porosity of the subsamples ranges from 0.11 to 0.18; the permeability of the subsamples varies from 16 to 122 mD, with the permeability of the whole sample being 73 mD. These ranges are even wider for the oil sand sample, which has porosity 0.26 and permeability 480 mD: the subsample porosity range is between 0.21 and 0.36 while the respective permeability falls between 278 and 1432 mD. However, when the permeability thus computed is plotted versus porosity, definite and fairly tight trends appear (Figure 19.4).

Such trends often emerge for other rock properties. Consider, for example a laboratory dataset that contains about 50 carbonate outcrop samples (Scotellaro *et al.*, 2008).

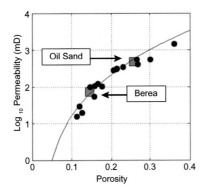

Figure 19.4 Computed permeability versus porosity for the Berea and oil sand samples and subsamples shown in Figure 19.3. Gray squares are for whole samples while black circles are for the subsamples. The curve is from the theoretical Kozeny–Carman equation (Mavko *et al.*, 2009). After Dvorkin *et al.* (2011).

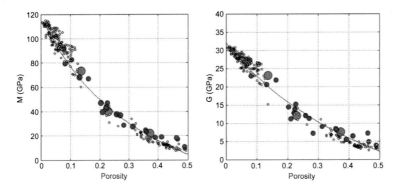

Figure 19.5 Carbonate dataset from Scotellaro *et al.* (2008). Small light symbols are laboratory measurements. Large gray circles are for computed properties of three digital samples. Smaller gray circles are for the subsamples of the three digital samples (eight per sample). The curves are from the stiff-sand model for pure calcite (after Dvorkin *et al.*, 2011).

The laboratory-measured compressional and shear moduli are plotted versus porosity in Figure 19.5. In the same figure we plot the elastic moduli computed on three digital samples as well as those computed on eight subsamples of each of the digital samples.

The two examples shown here, one for permeability and the other for the elastic moduli indicate that the physical properties computed on a few microscopic samples and their subsamples can form a trend that is close to the trend formed by a large number of laboratory experimental data, meaning that such trends persist in a wide scale range. In other words, such trends can be stationary versus scale. Moreover, a trend may be hidden inside a very small sample and extracted from a whole sample by subsampling it.

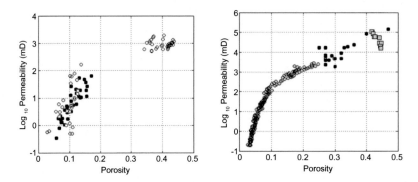

Figure 19.6 Left: Permeability and porosity computed on carbonate samples (black) compared to laboratory data from Scotellaro *et al.* (2008) shown as light symbols. Small light symbols are laboratory measurements. Right: Permeability and porosity computed on high-porosity oil sand samples compared to Fontainebleau sandstone dataset (Bourbie and Zinszner, 1985) and high-porosity Ottawa sand.

Hence, to make DRP useful, we need to produce trends rather than disparate data points, pretty much following the physical laboratory approach where a useful dataset contains measurements from a large number of samples of the same lithological category. One advantage of DRP over the physical laboratory is that the former allows for rapidly creating such trends from only a few digital samples. To validate DRP results, we have to compare trends rather than disparate data points.

19.3   More examples

The following case studies from Dvorkin *et al.* (2009) show how relations between porosity and permeability, porosity and the electrical formation factor, and porosity and velocity can be extracted from DRP experiments.

Figure 19.6 (left and right) shows permeability–porosity cross-plots obtained for carbonate samples and high-porosity oil sand, respectively. The first trend falls close to the relevant laboratory data while the second one falls upon the laboratory trend formed by high-porosity Ottawa sand and Fontainebleau sandstone. The electrical formation factor versus porosity computed for Fontainebleau sandstone and micritic carbonate is shown in Figure 19.7 and compared to laboratory data. Figure 19.8 shows an example of generating a formation factor versus porosity trend from two carbonate digital samples.

Figure 19.9 shows computed and measured velocity versus porosity for Fontainebleau sandstone, while Figure 19.10 shows digital velocity–porosity data for carbonate samples, together with laboratory data and a theoretical stiff-sand curve for calcite.

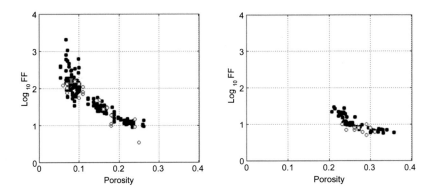

Figure 19.7 Electrical formation factor versus porosity for Fontainebleau sandstone (left) and micritic carbonate (right). Black symbols are for DRP results while light symbols are laboratory measurements. After Dvorkin *et al.* (2011).

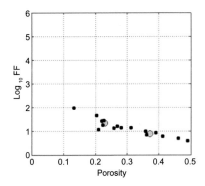

Figure 19.8 Electrical formation factor versus porosity for two digital carbonate samples (large gray circles) and their subsamples (black circles). After Dvorkin *et al.* (2011).

19.4 Multiphase flow

Tolke *et al.* (2010) discuss computer simulations of two-phase flow in the digital pore space. The time it takes to conduct such DRP experiments is orders of magnitude smaller than the time used in the physical laboratory. Moreover, such digital experiments are repeatable on exactly the same digital sample and are well suited for "what if" scenarios where the wettability and viscosity of the fluid phases can vary.

Figure 19.11 shows how the fractional volume flux of water in a decane/water system (defined as the ratio of the volume of water at the outlet of the sample per time unit to the total volume of all fluid phases) in digital pack of glass beads varies versus steady-state water saturation in the sample. Water in this example is the wetting fluid. The second example in the same figure is for water/oil flow where the ratio of the water/oil viscosity is 7/1 and 1/1. Figure 19.12 shows relative permeability curves obtained for a digital sandstone sample.

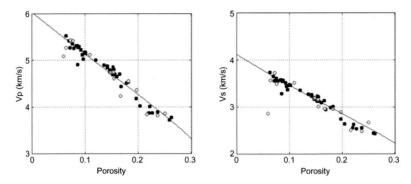

Figure 19.9 Velocity versus porosity for Fontainebleau sandstone. Black symbols are for DRP results while light symbols are laboratory measurements. Black curves are from the stiff-sand model for pure quartz. After Dvorkin *et al*. (2011).

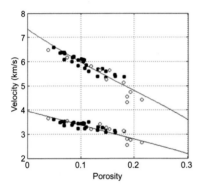

Figure 19.10 Velocity versus porosity for carbonate samples. Black symbols are for DRP results while light symbols are laboratory measurements. Black curves are from the stiff-sand model for pure calcite. After Dvorkin *et al*. (2011).

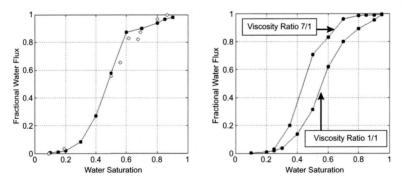

Figure 19.11 Flow in a glass bead pack. Left: Water/decane system. Black symbols are from the digital experiment and light symbols are physical experimental data. Right: Water/oil system with varying viscosity ratio as marked in the plot. After Tolke *et al*. (2010).

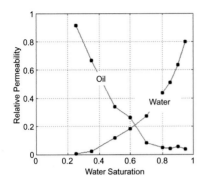

Figure 19.12 Flow in a digital sandstone sample. Relative permeability curves for water and oil with viscosity ratio 7/1 and surface tension 7 dyne/cm = 0.007 N/m. Tolke *et al.* (2010).

19.5 Conclusion

Computational rock physics does not diminish the importance of physical laboratory data, which will always serve as a benchmark and verification source. On the contrary, DRP can enrich physical data, especially so if used in the "what if" mode. Moreover, in some cases where only irregularly shaped and small rock samples are available (e.g., drill cuttings), DRP can serve as the only experimental tool.

Computational rock physics is uniquely designed for finding transforms between different rock properties: although it is virtually impossible to subsample a physical sample and consistently conduct the same laboratory experiments on each of the sub-samples, it is straightforward to accomplish this task on a computer. Following this, computational and analytical techniques can be used to ensure the utility of such trends in a range of scales. To paraphrase the famous saying, by using DRP we may be able "to see the rock in a grain of sand."

Appendix: Direct hydrocarbon indicator checklist

The following questionnaire is modified following Roden *et al.* (2005), Forrest *et al.* (2010), Roden *et al.* (2012), and Brown (2011).

Feasibility studies

(1) Can expected reservoir and hydrocarbon indicators, including those established for plausible variants of rock and pore fluid properties, be visible in ideal seismic data?

(2) Exactly what reservoir properties do we wish to model with the results from quantitative interpretation study: reservoir connectivity, net-to-gross, net pay, lithofacies, fluid saturation, fluid contacts, porosity, clay content?

(3) Are sufficient geological and petrophysical data available?

(4) Do the expected maximum/average ranges of the variables under examination (pore fluid, lithology, reservoir quality) produce clear effects on the acoustic and elastic impedance?

(5) What is the amplitude variation in the apparent water leg of the reservoir?

Anomaly and prospect

(6) Would you recognize this feature as a closure without the amplitude anomaly?

(7) Would you recognize this feature as a reservoir without the amplitude anomaly?

(8) Would this location be considered for drilling with the amplitude anomaly? (Especially important where amplitudes are expected from rock physics study but none are observed).

(9) Is the reflection from the suspected reservoir anomalous in amplitude?

(10) Where is the nearest discovery with the closure and formation?

(11) Where is the nearest discovery with the same formation with any closure type?

(12) Where is the nearest well with the reservoir quality rock in the target interval?

Data quality assessment

(13) Are seismic acquisition and processing parameters appropriate for the prospect?

(14) What is the primary seismic data used for DHI analysis (e.g., pre-stack migrated data)?

(15) What is the overall seismic image quality for amplitude analysis?

(16) Have relative amplitudes been preserved?

(17) What is the phase of the data?

(18) Are the data zero phase and known polarity?

(19) Does the data contain sufficient angle or offset for accurate AVO analysis?

(20) Are the stacking and migration velocities appropriate?

(21) How accurately is the angle of incidence known (velocity information)?

(22) Can the available seismic field data be used to predict the lateral changes of the relevant rock properties at the seismic scale?

(23) Is the vertical resolution of the seismic data sufficient for the problem under examination?

(24) Gathers: Are the pre-stack data properly processed (e.g., NMO corrected and noise attenuated)?

(25) Gathers: Can reflectors be clearly distinguished from random and coherent noise?

(26) Gathers: Are the reflectors properly flattened?

Seismic anomaly characteristics (phase, polarity, and shape)

(27) Do bright spots, dim spots, and/or phase change show the appropriate zero-phase character?

(28) Is the phase change (polarity reversal) visible?

(29) Is there one reflection from the top of the reservoir and one from the base?

(30) Do the top and base reflections exhibit natural pairing, dimming at the same point at the edge of the reservoir?

(31) If the reservoir is thick, are there significant reflections inside?

(32) Is there broadening of the reservoir reflections or a low-frequency shadow below?

Seismic anomaly characteristics (local change in amplitude)

(33) Amplitude versus background (A / B). Is the amplitude of the anomaly large relatively to the background?

(34) How consistent is the amplitude within mapped target area (on stacked data)?

(35) Are unexplained anomalies seen on stacked data outside the closure (within the same stratigraphic sequence)?

(36) Is the amplitude event unique?

Seismic anomaly characteristics (edge effects)

(37) Is down-dip conformance to closure present?

(38) Are the volumes in time or depth?

(39) Are the stratigraphic changes (such as channels) taken into account?

(40) Is there lateral conformance of structures based on far-offset or stacked data?

Seismic anomaly characteristics (flat spots)

(41) Flat spots can be observed when gross sand thickness is greater than tuning thickness. Hence, what is the tuning thickness?

(42) Is a flat spot visible and is it discrete?

(43) Is the flat spot perfectly flat or dipping consistently with gas velocity sag or tuning?

(44) If the flat spot does not conform with the structure, is it still consistent with the structure?

(45) Does the flat spot have the correct zero-phase character?

(46) Does the flat spot have the correct AVO response?

(47) Is the flat spot located at the down-dip limit of the anomaly?

(48) Is the phase change structurally consistent and at the same level as the flat spot?

(49) Do statistical cross-plotting techniques indicate a flat spot?

Seismic anomaly characteristics (data versus model)

(50) Do the well-based synthetic seismic traces match the actual seismic data?

(51) Do high-amplitude reflections have the character expected for hydrocarbons at this site?

(52) Excluding possible stacked pays, is the AVO effect anomalous compared to events above and below?

(53) Is the AVO effect anomalous compared to the same event outside the closure?

(54) How well defined is the background trend?

(55) How distinct is the anomaly from the background trend?

(56) Can the divergence from the background trend be explained by hydrocarbon substitution modeling?

Seismic anomaly characteristics (vertical and lateral context)

(57) Are multiple stacked indicators present at the same trap?
(58) Are there similar indicators at other parts of the closure?
(59) Is there velocity pull-down?
(60) Is there an amplitude and frequency shadow beneath the anomaly?
(61) Is there an anomaly in move-out-derived interval velocity?

Seismic analogues

(62) Have hydrocarbon indicators been proven nearby (true positive)?
(63) Have indicators been disproved nearby (false positive)?

Containment and preservation

(64) At the anomaly level, how confident are you in hydrocarbon trap preservation (e.g., late fault movement, breaching, or tilting)?
(65) What is the effective stress based on the sediment column height?

References

Aki, K. and Richards, P. G. (1980). *Quantitative Seismology: Theory and Methods*. W.H. Freeman and Co.

Anselmetti, F. S. and Eberly, G. P. (1997). Sonic velocity in carbonate sediments and rocks. In Palaz, I. and Marfurt, K. J., eds, *Carbonate Seismology, Geophysical Developments*. Tulsa, OK, USA: SEG, 53–74.

Arns, C. H., Knackstedt, M. A., Pinczewski, W. V. and Garboczi, E. J. (2002). Computation of linear elastic properties from microtomographic images: Methodology and agreement between theory and experiment, *Geophysics*, 67, 1396–1405, doi: 10.1190/1.1512785.

Athy, L. F. (1930). Density, porosity, and compaction of sedimentary rocks, *AAPG Bulletin*, 14, 1–24.

Avseth, P. (2000). *Combining rock physics and sedimentology for seismic reservoir characterization of North Sea turbidite systems*. Ph.D. thesis, Stanford University.

Avseth, P., Mukerji, T., and Mavko, G. (2005). *Quantitative Seismic Interpretation: Applying Rock Physics Tools to Reduce Interpretation Risk*. Cambridge University Press.

Avseth, P., Dvorkin, J., Mavko, G. and Rykkje, J. (2000). Rock physics diagnostic of North Sea sands: link between microstructure and seismic properties, *Geophysical Research Letters*, 27, 2761–2764, doi: 10.1029/1999GL008468.

Bachrach, R. and Avseth, P. (2008). Rock physics modeling of unconsolidated sands: accounting for nonuniform contacts and heterogeneous stress fields in the effective media approximation with applications to hydrocarbon exploration, *Geophysics*, 73, E197–E209.

Backus, G. F. (1962). Long-wave elastic anisotropy produced by horizontal layering, *Journal of Geophysical Research*, 67, 4427–4441, doi: 10.1190/1.1443207.

Baldwin, B. and Butler, C. O. (1985). Compaction curves, *AAPG Bulletin*, April, 69, 622–626.

Bangs, N. L., Sawyer, D. S. and Golovchenko, X. (1993). Free gas at the base of the gas hydrate zone in the vicinity of the Chile triple junction, *Geology*, 21, 905–908.

Batzle, M. and Wang, Z. (1992). Seismic properties of pore fluids, *Geophysics*, 57, 1396–1408, doi: 10.1190/1.1443207.

Batzle, M. L., Han, D.-H., and Hofmann, R. (2006). Fluid mobility and frequency-dependent seismic velocity – direct measurements, *Geophysics*, 71, N1–N9, doi: 10.1190/1.2159053.

Berryman, J. G. (1992). Single-scattering approximations for coefficients in Biot's equations of poroelasticity, *The Journal of the Acoustical Society of America*, 91, 551–571, doi: 10.1121/1.402518.

Blangy, J. P. (1992). *Integrated seismic lithologic interpretation: The petrophysical basis*. Ph.D. thesis, Stanford University.

Blatt, H., Middleton, G. and Murray, R. (1980). *Origin of Sedimentary Rocks*. Prentice-Hall, Inc.

Boggs, S. (1995). *Principles of Sedimentology and Stratigraphy*. Prentice-Hall, Inc.

Bosl, W., Dvorkin, J. and Nur, A. (1998). A study of porosity and permeability using a lattice Boltzmann simulation. *Geophysical Research Letters*, 25, 1475–1478, doi: 10.1029/98GL00859.

Bourbie, T. and Zinszner, B. (1985). Hydraulic and acoustic properties as a function of porosity in Fountainebleau sandstone. *Journal of Geophysical Research*, 90, 11524–11532, doi: 10.1029/JB090iB13p11524.

Bowers, G. L. (1995). Pore pressure estimation from velocity data: accounting for overpressure mechanisms besides undercompaction. *SPE Drilling and Completion*, SPE 27488, 515–530, doi: 10.2118/27488-PA.

Bowers, G. L. (2002). Detecting high overpressure. *The Leading Edge*, 21, 174–177, doi: 10.1190/1.1452608.

Box, G. E. P. and Draper, N. R. (1987). *Empirical Model-Building and Response Surfaces*, Wiley.

Box, R. and Lowrey, P. (2003). Reconciling sonic logs with check-shot surveys: stretching synthetic seismograms. *The Leading Edge*, 22, 510, doi: 10.1190/1.1587672.

Brie, A., Pampuri, F., Marsala, A. F. and O. Meazza, O. (1995). Shear sonic interpretation in gas bearing sands. *Proceedings of SPE Annual Technical Conference and Exhibition*, SPE 30595, 701–710, doi: 10.2118/30595-MS.

Brown, A. (2011). *Interpretation of Three-Dimensional Seismic Data*. SEG.

Cadoret, T. (1993). *Effet de la saturation eau/gas sur les proprietes acoustiques des roches*, Ph.D. thesis, University of Paris, VII.

Calvert, R. (2005). *Insights and Methods for 4D Reservoir Monitoring and Characterization*. SEG and EAGE.

Castagna, J. P., Batzle, M. L. and Eastwood, R. L. (1985). Relationships between compressional-wave and shear-wave velocities in clastic silicate rocks, *Geophysics*, 50, 571–581, doi: 10.1190/1.1441933.

Castagna, J. P., Batzle, M. L. and Kan, T. K. (1993). Rock physics – The link between rock properties and AVO response, in Offset-dependent reflectivity – Theory and practice of AVO analysis. In, Castagna, J. P. and M. Backus, eds, *Investigations in Geophysics, 8*. SEG, pp. 135–171.

Castagna, J. P., Swan, H. W. and Foster, D. J. (1998). Framework for AVO gradient and intercept interpretation, *Geophysics*, 63, 948–956, doi: 10.1190/1.1444406.

Castagna, J. P., Sun, S. and Siegfried, R. W. (2003). Instantaneous spectral analysis: detection of low-frequency shadows associated with hydrocarbons, *The Leading Edge*, 22, 120–127, doi: 10.1190/1.1559038.

Chatenever, A. and Calhoun, J. C. (1952). Visual examinations of fluid behavior in porous media – Part 1, *AIME Petroleum Transactions*, 195, 149–195, doi: 10.2118/135-G.

Chen, G., Matteucci, G., Fahmy, B. and Finn, C. (2008). Spectral-decomposition response to reservoir fluids from a deepwater West Africa reservoir, *Geophysics*, 73, 23–30, doi: 10.1190/1.2978337.

Connolly, P. (1999). Elastic impedance, *The Leading Edge*, 19, 438–452, doi: 10.1190/1.1438307.

Cordon, I., Dvorkin, J. and Mavko, G. (2006). Seismic reflections of gas hydrate from perturbational forward modeling, *Geophysics*, 71, F165–F171, doi: 10.1190/1.2356909.

Dai, J., Xu, H., Shyder, F. and Dutta, N. (2004). Detection and estimation of gas hydrates using rock physics and seismic inversion: examples from the northern deepwater Gulf of Mexico, *The Leading Edge*, 23, 60–66.

De Jager, J. (2012). Prospect evaluation and risk and volume assessment, Lecture notes, upublished.

Deutsch, C.V. and Journel, A. G, (1996). *GSLIB: Geostatistical software library and user's guide*, 2nd edn. Oxford University Press.

Dickey, P. (1992). La Cira-Infantas Field, Middle Magdalena Basin. In E. A. Beaumont and N. H. Foster, eds, *Structural Traps VII, AAPG Treatise of Petroleum Geology, Atlas for Oil and Gas Field*. AAPG, pp. 323–347.

Domenico, S.N. (1977). Elastic properties of unconsolidated porous sand reservoirs, *Geophysics*, 42, 1339–1368, doi: 10.1190/1.1440797.

Dutta, N. C. (1987). Fluid flow in low permeable porous media, in Migration of hydrocarbons in sedimentary basins. In B. Doligez, ed., *2nd IFP Exploration Research Conference*, Carcans, France, June 15–19. Editions Technip.

Dutta, N., Utech, R. and Shelander, D. (2010). Role of 3D seismic for quantitative shallow hazard assessment in deepwater sediments, *The Leading Edge*, 29, 930–942, doi: 10.1190/1.3480006.

Dvorkin, J. (2007). Self-similarity in rock physics, *The Leading Edge*, 26, 946–950, doi: 10.1190/1.2775996.

Dvorkin, J. (2008a). Yet another Vs equation, *Geophysics*, 73, E35–E39, doi: 10.1190/1.2792795.

Dvorkin, J. (2008b). The physics of 4D seismic, *Fort Worth Basin Oil and Gas Magazine*, October 2008, 33–36.

Dvorkin, J. (2008c). Can gas sand have a large Poisson's ratio?, *Geophysics*, 73, E51–E57, doi: 10.1190/1.2369900.

Dvorkin, J. (2008d). Seismic-scale rock physics of methane hydrates, *Fire in the Ice, DOE/NETL Methane Hydrate Newsletter*, Summer 2008, 13–17.

Dvorkin, J. (2009). Digital rock physics bridges scales of measurement, *E&P*, 82, 9, 31–35.

Dvorkin, J. and Alkhater, S. (2004). Pore fluid and porosity mapping from seismic, *First Break*, 22, 53–57, doi: 10.3997/1365–2397.2004003.

Dvorkin, J. and Brevik, I. (1999). Diagnosing high-porosity sandstones: strength and permeability from porosity and velocity, *Geophysics*, 64, 795–799, doi: 10.1190/1.1444589.

Dvorkin, J. and Cooper, R. (2005). The caveat of scale, *E&P*, 78, 10, 83–86.

Dvorkin, J. and Derzhi, N. (2013). Rules for upscaling for rock physics transforms: composites of randomly and independently drawn elements, *Geophysics*, 77, WA120–WA139, doi: 10.1190/geo2011–0268.1.

Dvorkin, J. and Gutierrez, M. (2001). *Textural Sorting Effect on Elastic Velocities, Part II: Elasticity of a Bimodal Grain Mixture*. SEG Technical Program Expanded Abstracts 2001, 1764–1767. Read more: http://library.seg.org/doi/abs/10.1190/1.1816466.

Dvorkin, J. and Gutierrez, M., 2002, Grain sorting, porosity, and elasticity, *Petrophysics*, 43, 3, 185–196.

Dvorkin, J. and Mavko, G. (2006). Modeling attenuation in reservoir and non-reservoir rock, *The Leading Edge*, 25, 194–197, doi: 10.1190/1.2172312.

Dvorkin, J. and Nur, A. (1996). Elasticity of high-porosity sandstones: theory for two North Sea datasets, *Geophysics*, 61, 1363–1370, doi: 10.1190/1.1444059.

Dvorkin, J. and Nur, A. (1998). Time-average equation revisited, *Geophysics*, 63, 460–464, doi: 10.1190/1.1444347.

Dvorkin, J. and Nur, A. (2009). Scale of experiment and rock physics trends, *The Leading Edge*, 28, 110–115, doi: 10.1190/1.3064155.

Dvorkin, J. and Uden, R. (2004). Seismic wave attenuation in a methane hydrate reservoir, *The Leading Edge*, 23, 730–734, doi: 10.1190/1.1786892.

Dvorkin, J. and Uden, R. (2006). The challenge of scale in seismic mapping of hydrate and solutions, *The Leading Edge*, 25, 637–642, doi: 10.1190/1.2202670.

Dvorkin, J., Mavko, G. and Nur, A. (1999). Overpressure detection from compressional- and shear-wave data, *Geophysical Research Letters*, 26, 3417–3420, doi: 10.1029/1999GL008382.

Dvorkin, J., Gutierrez, M. and Nur, A. (2002). On the universality of diagenetic trends, *The Leading Edge*, 21, 40–43.

Dvorkin, J., Nur, A., Uden, R. and Taner, T. (2003). Rock physics of a gas hydrate reservoir, *The Leading Edge*, 22, 842–847, doi: 10.1190/1.1614153.

Dvorkin, J., Walls, J., Uden, R., Carr, M., Smith, M. and Derzhi, N. (2004). Lithology substitution in fluvial sand, *The Leading Edge*, 23, 108–114, doi: 10.1190/1.1651452.

Dvorkin, J., Mavko, G. and Gurevich, B. (2007). Fluid substitution in shaley sediment using effective porosity, *Geophysics*, 72, O1–O8, doi: 10.1190/1.2565256.

Dvorkin, J., Armbruster, M., Baldwin, C., Fang, Q., Derzhi, N., Gomez, C., Nur, A. and Mu Y. (2008). The future of rock physics: computational methods versus lab testing, *First Break*, 26, 63–68, doi: 10.1029/1999GL008382.

Dvorkin, J., Derzhi, N., Fang, Q., Nur, A., Grader, A., Baldwin, C., Tono, H. and Diaz, E. (2009). From micro to reservoir scale: Permeability from digital experiments, *The Leading Edge*, 28, 1446–1453, doi: 10.1190/1.3272699.

Dvorkin, J., Derzhi, N., Diaz, E. and Fang, Q. (2011). Relevance of computational rock physics, *Geophysics*, 76, E141–E153.

Eastwood, J., Lebel, P., Dilay, A. and Blakeslee, S. (1994). Seismic monitoring of steam-based recovery of bitumen, *The Leading Edge*, 13, 242–251, doi: 10.1190/1.1437015.

Eaton, B. A. (1975). The equation for geopressured prediction from well logs, *Proceedings of Fall Meeting of the Society of Petroleum Engineers of AIME*, SPE 5544, doi: 10.2523/5544-MS.

Ebaid, H., Tura, A., Nasser, M., Hatchell, P., Smit, F., Payne, N., Herron, D., Stanley, D., Kaldy, J. and Barousse, C. (2008). First dual-vessel high-repeat GoM 4D shows development options at Holstein field, *SEG Expanded Abstracts*, doi: 10.1190/1.3064000.

Eberli, G. P., Baechle, G. T., Anselmetti, F. S. and Incze, M. L. (2003). Factors controlling elastic properties in carbonate sediments and rocks, *The Leading Edge*, 22, 654–660, doi: 10.1190/1.1599691.

Ebrom, D. (2004). The low-frequency gas shadow on seismic sections, *The Leading Edge*, 23, 772, doi: 10.1190/1.1786898.

Ecker, C., Dvorkin, J. and Nur, A. (2000). Estimating the amount of gas hydrate and free gas from marine seismic data, *Geophysics*, 65, 565–573.

Einsele, G., Ricken, W., and Seilacher, A., eds. (1991). *Cycles and Events in Stratigraphy*. Springer-Verlag.

Evejen, H. M. (1967). Outline of a system of refraction interpretation for monotonic increase of velocity with depth. In Musgrave, A. W., ed., *Seismic Refraction Prospecting*. SEG, p. 290.

Fabricius, I. L., Mavko, G., Mogensen, C. and Japsen, P. (2002). Elastic moduli of chalk as a reflection of porosity, sorting, and irreducible water saturation, *SEG Expanded Abstracts*, 1903–1906, doi: 10.1190/1.1817063.

Fabricius, I. L., Baechle, G. T. and Eberli, G. P. (2010). Elastic moduli of dry and water-saturated carbonates – effect of depositional texture porosity and permeability, *Geophysics*, 75, 65–78, doi: 10.1190/1.3374690.

Fahmy, W. (2006). *DHI/AVO Best Practices Methodology and Application*, SEG/AAPG 2006 Fall Distinguished Lecture.

Fahmy, W. A., Matteucci, G., Parks, J., Matheney, M. and Zhang, J. (2008). *Extending the Limits of Technology to Explore Below the DHI Floor; Successful Application of Spectral Decomposition to Delineate DHI's Previously Unseen on Seismic Data*. SEG Technical Program Expanded Abstracts 2008, 408–412.

Faust, L. Y. (1951). Seismic velocity as function of depth and geological time, *Geophysics*, 16, 192–206, doi: 10.1190/1.1437658.

Faust, L. Y. (1953). A velocity function including lithologic variation, *Geophysics*, 18, 271–288, doi: 10.1190/1.1437869.

Forrest, M., Roden, R. and Holeywell, R. (2010). Risking seismic amplitude anomaly prospects based on database trends, *The Leading Edge*, 29, 936–930, doi: 10.1190/1.3422455.

Fournier, F. and Borgomano, J. (2007). Geological significance of seismic reflections and imaging of reservoir architecture in the Malampaya gas field (Philippines), *AAPG Bulletin*, 92, 235–258, doi: 10.1306/10160606043.

Gal, D., Dvorkin, J. and Nur, A. (1998). A physical model for porosity reduction in sandstones, *Geophysics*, 63, 454–459, doi: 10.1190/1.1444346.

Gal, D., Dvorkin, J. and Nur, A. (1999). Elastic-wave velocities in sandstones with non-load-bearing clay, *GRL*, 26, 939–942.

Garboczi, E. J. and Day, A. R. (1995). An algorithm for computing the effective linear elastic properties of heterogeneous materials: three dimensional results for composites with equal phase Poisson's ratios, *Journal of the Mechanics and Physics of Solids*, 43, 1349–1362, doi: 10.1016/0022-5096(95)00050-S.

Gassmann, F. (1951). Elasticity of porous media: Uber die elastizitat poroser medien: Vierteljahrsschrift der Naturforschenden, *Gesellschaft*, 96, 1–23.

Ghaderi, A. and Landrø, M. (2009). Estimation of thickness and velocity changes of injected carbon dioxide layers from prestack time-lapse seismic data, *Geophysics*, 74, O17–O28, doi: 10.1190/1.3054659.

Ghosh, R. and Sen., M. (2012). Predicting subsurface CO_2 movement: from laboratory to field scale, *Geophysics*, 77, M27–M37, doi: 10.1190/geo2011-0224.1.

Giles, M. (1997). *Diagenesis: A Quantitative Perspective and Implications for Basin Modeling and Rock Property Prediction*. Kluwer Academic Publishers, p. 526.

Gommesen, L., Dons, T., Hansen, H. P., Jan Stammeijer, J. and Hatchell, P. (2007). 4D seismic signatures of North Sea chalk – the Dan field, *SEG Expanded Abstracts*, 2847–2851, doi: 10.1190/1.2793058.

Grana, D. and Della Rossa, E. (2010). Probabilistic petrophysical properties estimation integrating statistical rock physics with seismic inversion, *Geophysics*, 75, O21–O37, doi: 10.1190/1.3386676.

Grana, D., Mukerji, T., Dvorkin, J. and Mavko, G. (2012). Stochastic inversion of facies from seismic data based on sequential simulations and probability perturbation method, *Geophysics*, 77, M53–M72, doi: 10.1190/geo2011-0417.1.

Greenberg, M. L. and Castagna, J. P. (1992). Shear-wave velocity estimation in porous rocks: theoretical formulation, preliminary verification and applications, *Geophysical Prospecting*, 40, 195–209, doi: 10.1111/j.1365-2478.1992.tb00371.x.

Grotsch, J. and Mercadier, C. (1999). Integrated 3-D reservoir modelling based on 3-D seismic: the Tertiary Malampaya and Camago buildups, offshore Palawan, Philippines. *AAPG Bulletin*, 83, 1703–1728.

Grude, S., Dvorkin, J. and Landro, M. (2013). Rock physics estimation of cement type and impact on the permeability for the Snohvit Field, the Barents Sea, *SEG Expanded Abstract*.

Guerin, G. and Goldberg, D. (2002). Sonic waveform attenuation in gas-hydrate-bearing sediments from the Mallik 2L-38 research well, Mackenzie Delta, Canada, *Journal of Geophysical Research*, 107, 1029–1085, doi: 10.1029/2001JB000556.

Guerin, G., Goldberg, D. and Meltzer, A. (1999). Characterization of in-situ elastic properties of gas-hydrate-bearing sediments on the Blake Ridge, *JGR*, 104, 17781–17796.

Gutierrez, M. A. (2001). *Rock physics and 3-D seismic characterization of reservoir heterogeneities to improve recovery efficiency*. Ph.D. thesis, Stanford University.

Gutierrez, M. A. and Dvorkin, J. (2010). Rock physics workflows for exploration in frontier basins, *SEG Expanded Abstracts*, 2441–2446, doi: 10.1190/1.3513344.

Gutierrez, M. A., Braunsdorf, N. R. and Couzens, B. A. (2006). Calibration and ranking of pore-pressure prediction models, *The Leading Edge* 25, 1458–1460, doi: 10.1190/1.2369808.

Hackert, C. L. and Parra, J. O. (2004). Improving Q estimates from seismic reflection data using well-log-based localized spectral correction, *Geophysics*, 69, 1521–1529, doi: 10.1190/1.1836825.

Hamilton, E. L. (1972). Compressional-wave attenuation in marine sediments, *Geophysics*, 37, 620–646, doi: 10.1190/1.1440287.

Han, D.-H. (1986). *Effects of porosity and clay content on acoustic properties of sandstones and unconsolidated sediments*. Ph.D. thesis, Stanford University.

Hardage, B. A. (1985). *Vertical Seismic Profiling, Part A, Principles*, 2nd edn. Elsevier.

Hardage, B., Levey R., Pendleton, V., Simmons J. and Edson, R. (1994). A 3-D seismic case history evaluating fluvially deposited thin-bed reservoirs in a gas-producing property, *Geophysics*, 59, 1650–1665, doi: 10.1190/1.1443554.

Hashin, Z. and Shtrikman, S. (1963). A variational approach to the elastic behavior of multiphase materials, *Journal of Mechanics and Physics of Solids*, 33, 3125–3131, doi: 10.1016/0022-5096(63)90060-7.

Helgerud, M. (2001). *Wave speeds in gas hydrate and sediments containing gas hydrate: a laboratory and modeling study*, Ph.D. thesis, Stanford University.

Helgerud, M., Dvorkin, J., Nur, A., Sakai, A. and Collett, T. (1999). Elastic-wave velocity in marine sediments with gas hydrates: effective medium modeling, *GRL*, 26, 2021–2024.

Hill, R. (1952). The elastic behavior of crystalline aggregate, *Proceedings of the Physical Society, London*, A65, 349–354, doi: 10.1088/0370-1298/65/5/307.

Hilterman, F. (1989). Is AVO the seismic signature of rock properties?, *SEG Expanded Abstracts*, 559–562, doi: 10.1190/1.1889652.

Hilterman, F. (2001). *Seismic amplitude interpretation, SEG distinguished instructor short course*.

Hilterman, F. and Z. Zhou (2009). Pore-fluid quantification: Unconsolidated versus consolidated sediments, *SEG Expanded Abstracts*, 331–335, doi: 10.1190/1.3255549.

Holbrook, W. S., Hoskins, H., Wood, W. T., Stephen, R. A. and Lizarralde, D. (1996). Methane hydrate and free gas on the Blake Ridge from vertical seismic profiling, *Science*, 273, 1840–1843.

Hyndman, R. D. and Spence, G. D. (1992). A seismic study of methane hydrate marine bottom simulating reflectors, *JGR*, 97, 6683–6698.

Japsen, P. (1993). Influence of lithology and Neogene uplift on seismic velocities in Denmark; implications for depth conversion of maps, *AAPG Bulletin*, 77, 194–211.

Japsen, P. (1998). Regional velocity-depth anomalies, North Sea Chalk: a record of overpressure and Neogene uplift and erosion, *AAPG Bulletin*, 82, 2031–2074

Japsen, P., Mukerji, T. and Mavko, G. (2007). Constraints on velocity-depth trends from rock physics models, *Geophysical Prospecting*, 55, 135–154, doi: 10.1111/j.1365-2478.2007.00607.x.

Jizba, D. L. (1991). *Mechanical and acoustic properties of sandstones and shales*. Ph.D. dissertation, Stanford University.

Johnson, D. L. (2001). Theory of frequency dependent acoustics in patchy-saturated porous media, *The Journal of the Acoustical Society of America*, 110, 682–694, doi: 10.1121/1.1381021.

Kameda, A. and Dvorkin, J. (2004). To see a rock in a grain of sand, *The Leading Edge*, 23, 790–794, doi: 10.1190/1.1786904.

Katahara, K. (2003). Analysis of overpressure on the Gulf of Mexico Shelf, *Proceedings of Offshore Technology Conference*, OTC 15293, doi: 10.4043/15293-MS.

Keehm, Y., Mukerji, T. and Nur, A. (2001). Computational rock physics at the pore scale: Transport properties and diagenesis in realistic pore geometries, *The Leading Edge*, 20, 180–183, doi: 10.1190/1.1438904.

Kenter, J., Podladchikov, F., Reinders, M., Van der Gaast, S., Fouke, B. and Sonnenfeld, M. (1997). Parameters controlling sonic velocities in a mixed carbonate-siliciclastic Permian shelf-margin (upper San Andres formation, Last Chance Canyon, New Mexico), *Geophysics*, 64, 505–520, doi: 10.1190/1.1444161.

Klimentos, T. (1995). Attenuation of P- and S-waves as a method of distinguishing gas and condensate from oil and water, *Geophysics*, 60, 447–458, doi: 10.1190/1.1443782.

Klimentos, T. and McCann, C. (1990). Relationships among compressional wave attenuation, porosity, clay content, and permeability in sandstones, *Geophysics*, 55, 998–1014, doi: 10.1190/1.1442928.

Knackstedt, M. A., Arns, C. H. and Pinczewski, W. V. (2003). Velocity-porosity relationships, 1: Accurate velocity model for clean consolidated sandstones, *Geophysics*, 68, 1822–1834, doi: 10.1190/1.1635035.

Knight, R., Dvorkin, J. and Nur, A. (1998). Seismic signatures of partial saturation, *Geophysics*, 63, 132–138, doi: 10.1190/1.1887210.

Koesoemadinata, A.P. and McMechan, G.A. (2001). Empirical estimation of viscoelastic seismic parameters from petrophysical properties of sandstone, *Geophysics*, 66, 1457–1470, doi: 10.1190/1.1487091.

Krief, M., Garat, J., Stellingwerff, J. and Ventre, J. (1990). A petrophysical interpretation using the velocities of P and S waves (full-waveform sonic), *The Log Analyst*, 31, 355–369.

Krumbein, W. C. and Dacey, M. F. (1969). Markov chains and embedded Markov chains in geology: *Mathematical Geology*, 1 (1), 79–96, doi: 10.1007/BF02047072.

Kvamme, L. and Havskov, J. (1989). *Q* in southern Norway, *Bulletin of the Seismological Society of America*, 79, 1575–1588.

Kvenvolden, K. A. (1993). Gas hydrates as a potential energy resource – a review of their methane content. In *The Future of Energy Gases – U.S.G.S. Professional Paper 1570*, pp. 555–561.

Lancaster, A. and Whitcombe, D. (2000). Fast-track 'colored' inversion, *SEG Expanded Abstracts*, 1572–1575, doi: 10.1190/1.1815711.

Lander, R. H. and Walderhaug, O. (1999). Reservoir quality predictions through simulation of sandstones compaction and quartz cementation, *AAPG Bulletin*, 83, 433–449.

Latimer, R. B. (2011). Inversion and interpretation of impedance data. In Brown, A.R., ed., *Interpretation of Three-Dimensional Seismic*. SEG and AAPG.

Laverde, F. (1996). *Estratigrafia de alta resolucion de la seccion corazonada en el campo*, La Cira: Ecopetrol, Technical report, 37 p.

Leary, P., Henyey, T. and Li, Y. (1988). Fracture related reflectors in basement rock from vertical seismic profiling at Cajon Pass, *Geophysical Research Letters*, 15, 1057–1060, doi: 10.1029/GL015i009p01057.

Lebedev, M., Toms-Stewart, J., Clennell, B., Pervukhina, M., Shulakova, V., Paterson, L., Müller, T.M., Gurevich, B. and Wenzlau, F. (2009). Direct laboratory observation of patchy saturation and its effects on ultrasonic velocities, *The Leading Edge*, 28, 24–27, doi: 10.1190/1.3064142.

Lee, M. W. (2002). Biot-Gassmann theory for velocities of gas hydrate-bearing sediments, *Geophysics*, 67, 1711–1719.

Lee, M. W. (2006). A simple method of predicting *S*-wave velocity, *Geophysics*, 71, F161–F164, doi: 10.1190/1.2357833.

Li, J. and Dvorkin, J. (2012). Effects of fluid changes on seismic reflections: predicting amplitudes at gas reservoir directly from amplitudes at wet reservoir, *Geophysics*, 77, D129–D140, doi: 10.1190/geo2011-0331.1.

Lilwall, R. (1988). Regional mb:Ms, Lg/Pg amplitude ratios and Lg spectral ratios as criteria for distinguishing between earthquakes and explosions: A theoretical study, *Geophysical Journal*, 93, 137–147, doi: 10.1111/j.1365-246 X.1988.tb01393.x.

Lucet, N. (1989). *Vitesse et attenuation des ondes elastiques soniques et ultrasoniques dans ler roches sous pression de confinement* (Velocity and attenuation of elastic sonic and ultrasonic waves in rocks under confining pressure). Ph.D. thesis, University of Paris.

Lucia, F. J. (2007). *Carbonate Reservoir Characterization*, 2nd edn. Springer.

Marion, D. and Jizba, D. (1997). Acoustic properties of carbonate rocks: use in quantitative interpretation of sonic and seismic measurements. In Palaz, I. and Marfurt, K. J., eds, *Carbonate Seismology, Geophysical Developments*. SEG, pp. 75–94.

Marion, D., Mukerji, T. and Mavko, G. (1994). Scale effects on velocity dispersion: from ray to effective medium theories in stratified media, *Geophysics*, 59, 1613–1619, doi: 10.1190/1.1443550.

Marsden, D., Bush, M. D. and Sik Johng, D. (1995). Analytic velocity functions, *The Leading Edge*, 14, 775–782, doi: 10.1190/1.1437161.

Mavko, G. and Jizba, D. (1991). Estimating grain-scale fluid effects on velocity dispersion in rocks, *Geophysics*, 56, 1940–1949, doi: 10.1190/1.1443005.

Mavko, G., Chan, C. and Mukerji, T. (1995). Fluid substitution: Estimating changes in Vp without knowing Vs, *Geophysics*, 60, 1750–1755, doi: 10.1190/1.1443908.

Mavko, G., Mukerji, T. and Dvorkin, J. (2009). *The Rock Physics Handbook: Tools for Seismic Analysis of Porous Media*, Cambridge University Press.

Menezes, C. and Gosselin, O. (2006). From logs scale to reservoir scale: upscaling of the petroelastic model, *Proceedings of SPE Europec/EAGE Annual Conference and Exhibition*, SPE 100233, doi: 10.2523/100233-MS.

Miall, A. D. (1996). *The Geology of Fluvial Deposits: Sedimentary facies, basin analysis and petroleum geology*. Springer-Verlag.

Miall, A. D. (1997). *The Geology of Stratigraphic Sequences*. Springer-Verlag.

Miller, J. J., Lee, M. W. and von Huene, R. (1991). An analysis of a seismic reflection from the base of a gas hydrate zone, offshore Peru, *AAPG Bull.*, 75, 910–924.

Mindlin, R.D. (1949). Compliance of elastic bodies in contact, *Transactions ASME*, 71, A-259, doi: 10.1007/978-1-4613-8865-4_24.

Minshull, T. A., Singh, S. C. and Westbrook, G. K. (1994). Seismic velocity structure at a gas hydrate reflector, offshore western Colombia, from full waveform inversion, *JGR*, 99, 4715–4734.

Morales, L. G., Podesta, D. J., Hatfield, W. C., Tanner, H., Jones, S. H., Barker, M. H., O'Donoghue, D. J., Mohler, C. E., Dubois, E. P., Jacobs, C. and Goss, C. R. (1958). *General Geology and Oil Occurrences of the Middle Magdalena Valley, Colombia: Habitat of Oil Symposium*. American Association of Petroleum Geologists, pp. 641–695.

Mukerji, T., Jorstad, A., Avseth, P., Mavko, G. and Granli, J. R. (2001). Mapping lithofacies and pore-fluid probabilities in a North Sea reservoir: seismic inversions and statistical rock physics, *Geophysics*, 66, 988–1001, di: 10.1190/1.1487078.

Murphy, W. F. (1982). *Effects of microstructure and pore fluids on the acoustic properties of granular sedimentary materials*. Ph.D. thesis, Stanford University.

Nur, A. (1969). *Effects of stress and fluid inclusions on wave propagation in rock*. Ph.D. thesis, MIT.

O'Brien, J. (2004). Seismic amplitudes from low gas saturation sands, *The Leading Edge*, 23, 1236–1243, doi: 10.1190/leedff.23.1236_1.

Øren, P. E. and Bakke, S. (2003). Reconstruction of Berea sandstone and pore-scale modeling of wettability effects, *Journal of Petroleum Science and Engineering*, 39, 177–199, doi: 10.1016/S0920-4105(03)00062-7.

Osdal, B., Husby, O., Aronsen, H. A., Chen, N. and Alsos, T. (2006). Mapping the fluid front and pressure buildup using 4D data on Norne Field, *The Leading Edge*, 25, 1134–1141, doi: 10.1190/1.2349818.

Ostrander, W.J. (1984). Plane-wave reflection coefficients for gas sands at non-normal angles of incidence, *Geophysics*, 49, 1637–164, doi: 10.1190/1.1441571.

Paillet, F., Cheng, C. and Pennington, W. (1992). Acoustic waveform logging: advances in theory and application, *Log Analyst*, 33, 239–258.

Palaz, I. and Marfurt, K. J., eds, (1997). *Carbonate Seismology, Geophysical Developments*. SEG.

Pearson, C., Murphy, J. and Hermes, R. (1986). Acoustic and resistivity measurements on rock samples containing tetrahydrofuran hydrates: laboratory analogues to natural gas hydrate deposits, *JGR*, 91, 14132–14138.

Pickett, G. R. (1963). Acoustic character logs and their applications in formation evaluation, *Journal of Petroleum Technology*, 15, 650–667, doi: 10.2118/452-PA.

Pratt, R. G., Bauer, K. and Weber, M. (2003). Cross-hole waveform tomography velocity and attenuation images of arctic gas hydrates, *SEG Expanded Abstracts*, 2255–2258, doi: 10.1190/1.1817798.

Pride, S. R., Harris, J. M., Johnson, D. L., Mateeva, A., Nihei, K. T., Nowack, R. L., Rector, J. W., Spetzler, H., Wu, R., Yamamoto, T., Berryman, J. G. and Fehler, M. (2003). Permeability dependence of seismic amplitudes, *The Leading Edge*, 22, 518–525, doi: 10.1190/1.1587671.

Quan, Y. and Harris, J. M. (1997). Seismic attenuation tomography using the frequency shift method, *Geophysics*, 62, 895–905, doi: 10.1190/1.1444197.

Ramm, M. and Bjørlykke, K. (1994). Porosity/depth trends in reservoir sandstones; assessing the quantitative effects of varying pore-pressure, temperature history and mineralogy, Norwegian shelf data, *Clay Minerals*, 29, 475–490, doi: 10.1180/claymin.1994.029.4.07.

Raymer, L. L., Hunt, E. R. and Gardner, J. S. (1980). An improved sonic transit time-to-porosity transform, *Transactions of the Society of Professional Well Log Analysts*, 21st Annual Logging Symposium, Paper P.

Ren, H., Hilterman, F., Zhou, Z. and Dunn, M. (2006). AVO equation without velocity and density, *SEG Expanded Abstracts*, 239–243, doi: 10.1190/1.2370016.

Rider, M. (2002). *The Geological Interpretation of Well Logs*, 2nd edn. Whittles Publishing.

Rio P., Mukerji, T., Mavko, G. and Marion, D. (1996). Velocity dispersion and upscaling in a laboratory-simulated VSP, *Geophysics*, 61, 584–593, doi: 10.1190/1.1443984.

Roden, R., Forrest M., and Holeywell R., 2005, The impact of seismic amplitudes on prospect risk analysis, *The Leading Edge*, 24, 706–711, doi: 10.1190/1.1993262.

Roden, R., Forrest, M. and Holeywell, R. (2012). Relating seismic interpretation to reserve/resource calculations: insights from a DHI consortium, *The Leading Edge*, 31, 1066– 1074, doi: 10.1190/tle31091066.1.

Rose, P. (2001). Risk analysis and management of petroleum exploration ventures, *AAPG Methods in Exploration Series*, No. 12.

Ruiz, F. J. (2009). *Porous grain model and equivalent elastic medium approach for predicting effective elastic properties of sedimentary rocks*. Ph.D. thesis, Stanford University.

Russell, B. (1998). *Introduction to Seismic Inversion Methods*. SEG.

Rutherford, S. R. and Williams, R. H. (1989). Amplitude versus offset variations in gas sands, *Geophysics*, 54, 680–688, doi: 10.1190/1.1442696.

Sain, R. (2010). *Numerical simulation of pore-scale heterogeneity and its effects on elastic, electrical, and transport properties*. Ph.D. thesis, Stanford University.

Sakai, A. (1999). Velocity analysis of vertical seismic profiling (VSP) survey at Japex/JNOC/GSC Mallik 2L-38 gas hydrate research well, and related problems for estimating gas hydrate concentration, *GSC Bulletin*, 544, 323–340.

Sams, M. S. and Williamson, P. R. (1993). Backus averaging, scattering and drift, *Geophysical Prospecting*, 42, 541–564, doi: 10.1111/j.1365–2478.1994.tb00230.x.

Sayers, C. M. (2002). Stress-dependent elastic anisotropy of sandstones, *Geophysical Prospecting*, 50, 85–95, doi: 10.1046/j.1365–2478.2002.00289.x.

Scholl, D. W. and Hart, P. E. (1993), Velocity and Amplitude Structures on Seismic-Reflection Profiles–Possible Massive Gas-Hydrate Deposits and Underlying Gas in *The Future of Energy Gases*, ed. D. G. Howell, pp. 331–351.

Schon, J. H. (2004). *Physical Properties of Rocks: Fundamentals and Principles of Petrophysics*, Elsevier.

Scotellaro, C., Vanorio, T. and Mavko, G. (2008). The effect of mineral composition and pressure on carbonate rocks, *SEG Expanded Abstracts*, 1684–1689, doi: 10.1190/1.2792818.

Sen, M. and Stoffa, P. L. (2013). *Global Optimization Methods in Geophysical Inversion*, 2nd edn. Elsevier.

Sharp, B., DesAutels, D., Powers, G., Young, R., Foster, S., Diaz, E. and Dvorkin, J. (2009). Capturing digital rock properties for reservoir modeling, *World Oil*, 230, 10, 67–68.

Sheriff, R. and Geldart, L. (1995). *Exploration Seismology*. Cambridge University Press.

Shuey, R. T. (1985). A simplification of the Zoeppritz equations, *Geophysics*, 50, 619–624, doi: 10.1190/1.1441936.

Slotnick, M. M. (1936). On seismic computations with applications, *Geophysics*, 1, 9–22, doi: 10.1190/1.1437084.

Spencer, J. W., Cates, M. E. and Thompson, D. D. (1994). Frame moduli of unconsolidated sands and sandstones, *Geophysics*, 59, 1352–1361, doi: 10.1190/1.1443694.

Strandenes, S. (1991). *Rock physics analysis of the Brent Group Reservoir in the Oseberg Field: Stanford Rock Physics and Borehole Geophysics Project*, special volume.

Su, Y., Tao, Y., Wang, T., Chen, G. and Li, J. (2010). AVO attributes interpretation and identification of lithological traps by prestack elastic parameters inversion – a case study in K Block, South Turgay Basin, *SEG Expanded Abstract*, 439–443, doi: 10.1190/1.3513794.

Taner, M. T., Koehler, F. and Sheriff, R. E. (1979). Complex seismic trace analysis, *Geophysics*, 44, 1041–1063, doi: 10.1190/1.1440994.

Tarantola, A. (2005). *Inverse Problem Theory*. SIAM.

Timur, A. (1968). An investigation of permeability, porosity, and residual water saturation relationships for sandstone reservoirs: *The Log Analyst*, 9, 4, 8–17.

Tolke, J., Baldwin, C., Mu, Y., Derzhi, N., Fang, Q., Grader, A. and Dvorkin, J. (2010). Computer simulations of fluid flow in sediment: From images to permeability, *The Leading Edge*, 29, 68–74, doi: 10.1190/1.3284055.

Toms, J., Muller, T. M., Cizc, R. and Gurevich, B. (2006). Comparative review of theoretical models for elastic wave attenuation and dispersion in partially saturated rocks, *Soil Dynamics and Earthquake Engineering*, 26, 548–565, doi: 10.1016/j.soildyn.2006.01.008.

Trani, M., Arts, R., Leeuwenburgh, O. and Brouwer, J. (2011). Estimation of changes in saturation and pressure from 4D seismic AVO and time-shift analysis, *Geophysics*, 76, C1–C17, doi: 10.1190/1.3549756.

Vanorio, T. and Mavko, G. (2011). Laboratory measurements of the acoustic and transport properties of carbonate rocks and their link with the amount of microcrystalline matrix, *Geophysics*, 76, E105–E115. doi: 10.1190/1.3580632.

Vanorio, T., Scotellaro, C. and Mavko, G. (2008). The effect of chemical processes and mineral composition on the acoustic properties of carbonate rocks, *The Leading Edge*, 27, 1040–1048, doi: 10.1190/1.2967558.

Vanorio, T., Nur, A. and Ebert, Y. (2011). Rock physics analysis and time-lapse rock imaging of geochemical effects due to the injection of CO_2 into reservoir rocks, *Geophysics*, 76, O23–O33, doi: 10.1190/geo2010-0390.1.

Vasquez, G. F., Dillon, L. D., Varela, C. L., Neto, G. S., Velloso, R. Q. and Nunes, C. F. (2004). Elastic log editing and alternative invasion correction methods, *The Leading Edge*, 23, 20–25, doi: 10.1190/1.1645452.

Vernik, L., Fisher, D. and Bahret, S. (2002). Estimation of net-to-gross from P and S impedance in deepwater turbidites, *The Leading Edge*, 21, 380–387, doi: 10.1190/1.1471602.

Walls, J., Dvorkin, J. and Smith, B. (1998). Modeling seismic velocity in Ekofisk chalk, *SEG Expanded Abstracts*, 1016–1019, doi: 10.1190/1.1820055.

Wang, Z. (1988). *Wave velocities in hydrocarbons and hydrocarbon saturated rocks – with application to EOR monitoring*. Ph.D. thesis, Stanford University.

Wang, Z. (1997). Seismic properties of carbonate rocks. In *Carbonate Seismology, Geophysical Developments*, Palaz, I. and Marfurt, K. J., eds. SEG, pp. 29–52.

Wang, Z. (2000). Velocity-density relationships in sedimentary rocks. In Wang, Z., Nur, A. and Ebrom, D. A., eds, *Seismic and Acoustic Velocities in Reservoir Rocks, Recent Developments* (Geophysics Reprint Series 19), SEG, pp. 256–268.

Waters K. H. (1992). *Reflection Seismology: A tool for energy resource exploration*, 3rd edn. Krieger.

White, J. E. (1983). *Underground Sound: Application of seismic waves*. Elsevier.

Williams, D. M. (1990). The acoustic log hydrocarbon indicator, SPWLA 31st Logging Symposium, Paper W.

Winkler, K. (1979). *The effects of pore fluids and frictional sliding on seismic attenuation*. Ph.D. thesis, Stanford University.

Wood, A. W. (1955). *A Textbook of Sound*. MacMillan.

Wood, W. T., Stoffa, P. L. and Shipley, T. H. (1994). Quantitative detection of methane hydrate through high-resolution seismic velocity analysis, *Journal of Geophysical Research*, 99, 9681–9695.

Wood, W. T., Holbrook, W. S. and Hoskins, H. (2000). In situ measurements of *P*-wave attenuation in the methane hydrate- and gas-bearing sediments of the Blake Ridge. In Paull, C. K., Matsumoto, R., Wallace, P. J. and Dillon, W. P., eds, *Proceedings of the Ocean Drilling Program, Scientific Results*, 164, 265–272.

Wyllie, M. R. J., Gregory, A. R. and Gardner, G. H. F. (1956). Elastic wave velocities in heterogeneous and porous media. *Geophysics*, 21, 41–70.

Yilmaz, O. (2001), *Seismic Data Analysis*. SEG.

Yin, H. (1992). *Acoustic velocity and attenuation of rocks: Isotropy, intrinsic anisotropy, and stress-induced anisotropy*. Ph.D. thesis, Stanford University.

Zhou, Z. and Hilterman, F. (2010). A comparison between methods that discriminate fluid content in unconsolidated sandstone reservoirs, *Geophysics*, 75, B47–B58, doi: 10.1190/1.3253153.

Zhou, Z., Hilterman, F. and Ren, H. (2006). Stringent assumptions necessary for pore-fluid estimation, *SEG Expanded Abstracts*, 244–248, doi: 10.1190/1.2370027.

Zimmer, M. A. (2003). *Seismic velocities in unconsolidated sands: measurements of pressure, sorting, and compaction effects*. Ph.D. thesis, Stanford University.

Zoeppritz, K. (1919). Erdbebenwellen VIIIB, On the reflection and propagation of seismic waves, *Gottinger Nachrichten*, I, 66–84.

Index